Media and The City

Media and The City:
Urbanism, Technology and Communication

Edited by

Simone Tosoni, Matteo Tarantino
and Chiara Giaccardi

CAMBRIDGE
SCHOLARS
PUBLISHING

Media and The City: Urbanism, Technology and Communication,
Edited by Simone Tosoni, Matteo Tarantino and Chiara Giaccardi

This book first published 2013

Cambridge Scholars Publishing

12 Back Chapman Street, Newcastle upon Tyne, NE6 2XX, UK

British Library Cataloguing in Publication Data
A catalogue record for this book is available from the British Library

ISBN (10): 1-4438-4943-X, ISBN (13): 978-1-4438-4943-2

TABLE OF CONTENTS

FOREWORD

WHATEVER HAPPENED TO *FLÂNERIE*? ON SOME THEORETICAL IMPLICATIONS OF THE MEDIA/CITY NEXUS

CHIARA GIACCARDI

To establish a focus on media and the city, as the following chapters do, means to cut a space of observation in which crucial processes occur, crossing and co-shaping each other, thus potentially fostering new perspectives on several important theoretical questions, from the social shaping of technology (McKenzie & Wajcman, 1985) to the post-media (or hyper-media) condition (Krauss, 2006), from participatory citizenship (Dahlgren, 1995) to new forms of aestheticisation of everyday life (Jameson, 1991), from the reshaping of mobility to the new regimes of perception and many others.

As a foreword to the first volume collecting the work of the ECREA[1] Temporary Working Group "Media & the City", which I have the honour to chair until 2014, I will briefly address a couple of those questions, namely the reconfiguration of sensitivity in the digital age and the new socio-aesthetic processes at play in the mixed environment of contemporary cities.

1. Reconfigurations

Media studies owe Marshall McLuhan (McLuhan, 1964) the presence of an analytical focus on the reconfigurations facilitated by media, and probably few places display this spectacle better than contemporary cities. In the nexus between media and the city we can spot at least two crucial anthropological aspects: our ability to communicate through shared,

[1] European Communication Research and Education Association: see www.ecrea.eu.

complex codes and our ability to build forms of settlement that are both permanent and susceptible to transformation.

Urban remediation (Bolter & Grusin, 1999) fosters new forms of "mixed lives" (Turkle, 2011), continuously renegotiating the boundaries between what is "analogue" and what is "digital"–which in turn gives rise to new poetics of space, involving new languages and, possibly, new contents.

A preliminary observation is needed: perceived space and "acted" space are not entirely distinguishable experiences nowadays (as testified by the popularity of perspectives and concepts such as the sensorimotor paradigm, enactive perception and haptic eye[2]). I am not completely sure they have ever been distinguishable, but the ever-growing penetration of media within the urban fabric certainly fosters new forms of intertwining. Not only because media are more and more explicitly integrated into the urban fabric as screens or other forms of information supply–according to the post-media view that acknowledges the new character of our mixed and convergent era (Krauss, 2006). Maybe there is something even more fundamental at core level: again drawing on McLuhan, the main reason is that media *transform the sensorium*–that is, the paradigm of our sensorial experience, or, as Ong (1967) defined it, "The patterned, patterning, coordinated world of the sense experience; the entire sensory apparatus as an operational complex". Changes in perceived space, therefore, open up new possibilities for action.

Let me briefly sketch some of the ways ICTs change the sensorium. First of all, they enable a much greater amount of data to impact our senses: "augmented reality" is a product of this "augmented sensitivity". In turn, this augmented sensitivity appears to be fostering a shift of the sensorium itself. From this perspective, the sensorium is shifting from a paradigm of well-distinct, specialized senses towards an indistinct paradigm of mutually-translating senses dominated by touch (which McLuhan correctly identified as the "interplay of senses") (McLuhan, 1996). From this crossmodal spatial experience, augmented reality technologies can be read precisely as enabling users to "touch" the surface of things: our eye (extended in our camera, cursor, mobile phone camera etc.) "touches" the space and data comes out of it. The prominence of

[2] On enactive perception and sensorimotor paradigm, see among the others O'Regan & Noe, 2001; on the haptic eye, see Merleau-Ponty, 2006.

tactility is incarnated even more explicitly as the driving design force of the majority of contemporary user interfaces. We want to touch stuff and see things happen (which is, incidentally, a working definition of "magic"): we touch to see, we touch to listen, we touch to remember.

This paradigmatic change fosters a new sense of reciprocity through the "always-on" status of contemporary media. Being in a state of what James E. Katz calls "perpetual contact" (Katz, 2004) not only enables but somehow *forces* us to be "in touch" with our social graph, and indeed the success of social networking utilities can be interpreted as a response to a growing effective *need* for such tools.

The augmentation of our sensitivity is ripe with consequences. New forms of resistance to the "strategies" of power (as De Certeau called them) are opened by this reconfiguration of the sensorium. Yet along with (and inseparable from) these new possibilities come new possible forms of discipline of urban bodies and discourses. An increased paradigm of choices runs the risk of imploding in hetero-direction and, as such, of fostering conformism.

Which leads us to our second observation: the city has *always* been a medium. Hyper-mediality (the condition in which the boundaries between media and environment become blurred) is only the latest development of a long-running process. This has been discussed at length, of course, but is particularly evident to me as an Italian. European cities have grown historically around their churches. Entire cosmologies, built around the duality of the sacred and the profane, have been vehiculated through spatial relationships. Not only the church itself is a space featuring the utmost density of codes, but consider its relationship with its square and bell tower; or, again, the geometrical–and symbolic–relationship between the bell tower and the other buildings. Or the inclusive audiospace described by the bells, with their ability to organise bodies and goods in the urban space, and to signal the cyclic ceremonial coincidence between sacred time *and* sacred space which was (and to some extent still is) the central axis of city life (Corbin, 1994; Illich, 2010).

Examples could go on indefinitely; yet the main point here is that cities have always communicated–and have done so by means of *immersion*, that is by establishing common, shared and symbolically dense spaces of experience to be embedded in through the sensorium. The implications of this space were of utmost importance: for instance, in defining and identifying the boundaries of identity, while emarginating "the stranger" as the individual showing a lack of adjustment to its specific affordances.

There is certainly nothing new in the ability of the city to communicate. However, what changes is the *remediation* of the urban space, which entails a *transformation of the forms of immersion* offered by the city, and therefore of the possible contents being communicated. Augmented Reality technologies effectively cover urban space with layers of information whose configurations change according to each user. While the individual dimension of the sensorium has always impacted the experience of immersion (for example, by separating "strangers" from "non-strangers"), the technology owned by the individual now impacts and diversifies the experience of immersion. Does the "shared space of experience" remain a shared, common space when the access to its significance is increasingly related to the possession of specific hardware/software combinations?

Conversely, on the "democratization" side, the possibility of accessing and decoding specific meanings today can generally bypass the requirement of specific codified knowledge (of "knowing about" the city) and can be enacted by anyone. The layers of history, architecture and language that have been part of urban space have been peeled back, remediated, digitised and piled back on, for anyone to access. By waving my smartphone around with an app such as Layar, or any other augmented reality application, I am accessing an (arbitrarily) translated, edited but otherwise *functional* version of those layers of knowledge that make up immersion. My technological gear (which is a function of my capital, positioning and social graph) potentially renders much of my knowledge about the city irrelevant, at the same time supplying me with copious amounts of the same knowledge.

In this sense the city becomes (or pretends to become, or is perceived as becoming, or is marketed as becoming) *transparent*, exposed. The decoding of its layers of meaning becomes semi-automatic. Yet is a poetry of space possible in such a transparent space? Stripped of its opacity, the city bares all of its stories for everyone to see: what remains is a naked city with a superimposed (though hyper-dense and ever-growing) informational layer. In other words: can a "naked city" be inspirational? As far as the poetic of space is concerned, one of the transformations at stake is related to the remediation of *flânerie*.

2. *Flânerie* and sociability: from inclusive aesthetics to participation

It was Georg Simmel who saw existential detachment as a product of the city; and detachment is a precondition for *flânerie*. *Flânerie* could only develop in the city. "Look but don't touch" is the *flâneur's* motto. And losing oneself within the urban space, effectively becoming *driven by urban* space, experiencing visual immersion at its foremost, has often been defined (first by Benjamin, of course) as one of the quintessential "modern" spatial practices. Walking around the city as a *flâneur*, Benjamin, Baudelaire and the Situationists would argue, is the best way to let urban space tell its stories–that is, to unveil its layer of information to our sensitivity. And by moving in a deliberately purposeless way, one can escape the functional logics of spatial arrangement (that is, for most of this thinkers, the logics of capital) as much as possible. Of course the logic of capital soon appropriated *flânerie* by replacing the arcade with the department store first and the shopping mall later–where an empty (yet perfectly functional) proxy of *flânerie* has since been offered at the price of consumption, as Bauman, among the others, poignantly noticed (Bauman & Lyon, 2012).

Flânerie is one of the fascinating paradoxes of modernity: one of the paramount acts of urban freedom (a subtraction from the capitalist logics regulating the city) is letting the metropolis itself (the paradigmatic structure of modernity) appropriate the body and the senses, and by doing so, in a thrilling reversal, effectively (if momentarily) re-appropriate city space.

Yet–what happens to remediated *flânerie* in the kind of urban space I just described, which is at the same time intensively shared and highly personal? I don't mean to discuss "new" forms of *flânerie*–that has been done ad nauseam: moviegoing, Disneyland, hypertext, TV zapping, all those have been described as new forms of *flânerie* at some point. Instead, what I am interested in is: what happens to the experience of "letting urban space drive our body"–and to the immersive communication–when our sensitivity is augmented?

I would argue that the layers of information about urban space accessed by augmented sensitivity effectively exasperate the *functionalisation* of the same spaces. Data has meaning only when queried: and queries are formulated according to functional requirements. The layers of information are often experienced as part of consumption practices of various kinds, including cultural consumption, tourism and entertainment.

One last note on *flânerie*. For Baudelaire, Benjamin and Debord, the point may not have been the "authenticity" of the experience of *flânerie*, on which most commentators have been focusing. The point may have been "immersion as key to inspiration": the city as the engine of further processes of signification and ποιέω (*poiesis*). In this sense, augmented reality, precisely because it is entangled in a layer of (more or less forced) sociability, offers great and yet under-explored possibilities: Foursquare check-ins, Facebook statuses, Tweets, may–*when seen from above*–be the verses of a kind of urban poetry far beyond anything produced so far. Immense amounts of information, stories and tales on, about and around urban space are produced every day by each and every one of us (bar those who are offline).

Of course, new aesthetics are needed to interpret such works, but the point here is that *flânerie* in itself may have changed beyond recognition, yet its purpose (to draw inspiration from the layers of information offered by the urban space) may have remained vital. The "naked city" might be a trick of the light for those of us who grew up in spaces to some extent closer to Baudelaire's than that of Foursquare creator Naveen Selvadurai.

Moreover, our contemporary, touch-intensive sensorium is by and large incompatible with Baudelaire's and Benjamin's sight-dominated paradigm: our experience of space is more and more dominated by a requirement of "touch". According to McLuhan, in fact, "it begins to be evident that 'touch' is not skin, but the interplay of senses, and 'keeping in touch' or 'getting in touch' is a matter of fruitful meeting of the senses, of sight translated into sound and sound into movement, and taste, and smell" (McLuhan, 1996, p. 60)

And what about sociability, in the era of social networks, flash mobs, geolocalization? As audio-tactile perception becomes dominant (along with its imperative of "being together"[3]), urban space reconfigures itself to accommodate for it, extending itself as a haptic, audiotactile environment. Yet what is gained in terms of complexification and stratification of space is lost in the effective possibility of withdrawing oneself from consumption

[3] According to McLuhan, "acoustic space has no centre and no margins", while tactility is favoured by the simultaneity of mixed stimula. Audio-tactile involving and inclusive environments harkens back to the tribal era, overcoming detachment, fragmentation, individualism of modernity (McLuhan, 1969).

and other solicitations. We are inside this space whether we like it or not, and withdrawal is becoming ever more expensive (to the point of being impossible for most of us). Among such solicitations is a sociability that becomes imperative: do things together, see things together, join, share, are the verbs of the day. *Flânerie* was a solitary experience, yet now deliberate solitude has become a luxury for many of us.

Social participation is a topical issue today, and is rather ambivalent. On one hand, any technological determinism should be rejected, be it in its enthusiastic or pessimistic version, as technology cannot "produce" participation *per se:* therefore, one issue to be addressed is certainly how we can render our cities more social rather than simply more high-tech.

Another important issue may be how to exceed the narrow but, up to this moment, dominant commercial scope of geolocation apps; in other words, whether mobile and location-based apps can be used behind personalized consumption and sharing preferences with an in-group of like-minded people; how new networked publics can be activated thanks to digital media, beyond top-down or bottom-up, but peer-to-peer; how to design interventions where individual use does not deplete the commons but instead adds value to the whole, moving towards a condition of "augmented deliberation"; how digital media can help to strengthen the sense of belonging and commitment to locality, that is, citizen engagement with collective urban issues and the power to act on them. The crucial words here are inclusiveness, access and agency: conceptual shifts in the notion of dwelling from "possession" to the right to act, collectively, for common goals. Participation can certainly take a step beyond crowdsourcing existing issues, where people only have a signalling role and/or a role as generators of ideas, but their right or capacity to act remains limited.

Moreover, another important issue is related to how ICTs function in the management of the struggle over the meaning of space (when contested among different groups), through participation, citizen journalism, coordination and so on. Then contiguity, transitivity and the mutual shaping of digital and material environment certainly set up new conditions for agency.

3. Building analytical frameworks for the media/city nexus

In any case, a new, interdisciplinary analytical framework is much needed in order to understand the city as a permanent workshop of innovative processes and as a magnifier to observe contemporary life. This approach should attempt to bring together aesthetic, sociological, anthropological, technological, historical and geographical sensibilities.

This realisation was the starting point of the "Media & the City" Temporary Working Group, established in mid-2011 within ECREA. The group's key priority is to foster a productive and empirically grounded dialogue among different perspectives, frames, competences, experiences and projects; moreover it has been establishing connections with other research and institutional bodies with similar interests.

The Working Group has grown around a core interest particularly suitable for this cooperative and interdisciplinary effort. This core is composed of three main focuses, namely:

a) *Media representations of urban space* and of related social processes;

b) *Cities as spaces for media usage and social practices*, and the influence of media in the experience of cities (including: geo-location, geo-annotation, new public and private spaces);

c) The *presence of media in the urban contexts* (including: new forms of architecture, the impact of security technologies, new forms of interaction with city spaces…).

While enthusiastic views on the possibilities of buildings to be "transformed from enclosed shelters into open environment" abound (as Paola Antonelli, senior curator in the Department of Architecture and Design at Moma NY once said[4]), on the other side there is a growing awareness that technological progress always brings formal innovation, which starts as creative flair, but may soon lapse into routine. Moreover, that exhibition has a dark side called surveillance (Bauman & Lyon, 2012).

The enthusiastic view of cities as "setting up a path that is transforming them into information parkour and enriching our lives with emotions, motion, direction, depth, and freedom"[5] is balanced by the new concern with tracking, exposition, transparency and new vulnerabilities (one of the

---

[4] In *Talk to me. Design and the communication between people and objects*, Exibition Catalogue, MOMA, New York, 2011, p. 9.

[5] Ibidem.

last issues being the uses of drones for urban surveillance). The ways in which attitudes and practices are changing under conditions of intensifying surveillance, and the way in which people comply with, negotiate or resist surveillance today are certainly among the emerging topics in the study of the media/city nexus.

It seems to me that one of the main goals of projects such as Media & the City is precisely that of engaging and disarming techno-utopian views of the relationship between ICTs and the city, fostering a critical and empirically grounded approach to what is actually happening instead.

References

Bauman, Z., & Lyon, D. (2012). *Liquid Surveillance*. Cambridge, MA: Polity Press.

Bolter, J., & Grusin, R. (1999). *Remediation*. Cambridge, MA: MIT Press.

Katz, J., & Aakhus, M. A. (2004). *Perpetual Contact.* Cambridge, MA: Cambridge University Press.

Corbin, A. (1994). *Les cloches de la terre*. Paris: Flammarion.

Dahlgren, P. (1995). *Television and the Public Sphere: Citizenship, democracy and the media*. London: Sage.

Illich, I. (2010). *I fiumi a Nord del futuro*. Macerata: Quodlibet.

Jameson, F. (1991). *Postmodernism, or the cultural logic of late capitalism*. Durham: Duke University Press.

Krauss, R. (2006). Two moments from the post-medium condition. *MIT Press Journal*, Spring, 55-62.

Lefebvre, H. (1991) [1974]. *The production of space*. Oxford: Wiley-Blackwell.

McLuhan, M. (1996) [1964]. *Understanding Media*. Boston: MIT Press.

—. (1969). Playboy Interview: Marshall McLuhan. *Playboy*, March, pp. 26-27, 45. 55-56, 61, 63.

Merleau-Ponty, M. (2006) [1964]. *L'oeil et l'ésprit*. Paris: Folio.

MacKenzie, D., & Wajcman, J. (Eds.). (1985). *The social shaping of technology*. Buckingham: Open University Press.

O'Regan, J. K., & Noe, A. (2001). A sensorimotor account of vision and visual consciousness. *Behavioural and Brain Sciences*, 24(5), 939-1031.

Ong, W. (1967). *The presence of the world.* New Haven: Yale University Press.

Turkle, S. (2011). *Alone Together*. New York: Basic Books.

Wellmann, B., & Rainie, L. (2012). *Networked*. Boston, MA: MIT Press.

Zukin, S. (2010). *Naked City*. New York: Oxford University Press.

Introduction

Towards a New Complexity: Reasons for Media and The City

Simone Tosoni and Matteo Tarantino

As two inherently "human" products–ambiguously *in-between* structure and agency, *langue* and *parole*–both cities and communication have been an endless source of fascination for human beings, including social science scholars. What's more, modernity has been pushing cities and communications (or more specifically, media) closer and closer to the centre of our lives, thus increasing the interest for an analysis of their intersection.

Walter Benjamin and George Simmel could be considered among the forefathers of this interest in how the modern metropolis communicates and is communicated (McQuire, 2008). Although their approach to communication was obviously not systematic, it is impossible not to be fascinated by the precognitive reflections of Benjamin on how everyday life became common knowledge in the new urban space (the "glass house" which he addressed quite enthusiastically) or by Simmel's analyses on how the city restructures perception.

Both authors were concerned with epochal transformations. This view of cities restructuring communication and sociality, or, conversely, of media transforming the very nature of urban space (often by making it null and void, or unnecessary) has been a strong current in social sciences for many years. Be it blissful transcendence or apocalypse, this view of the media/city relationship as macro-transformation has produced a wealth of works in the last decades. This is especially true since the transformations brought forward by post-fordism and neoliberalism have become manifest in the 1980s and 1990s. Indeed, the spatio-temporal changes related to what has come to be known as "globalization" were too large, urgent and exciting not to attract and orient scholarly reflection. Some generalizations and abstractions were an acceptable price to pay to come to grips with such large-scale phenomena. So cities became nodes, (e.g. Sassen, 1994; Castells, 1996; Borja & Castells, 1997), disappeared, became digital (Mitchell, 1996; Graham, 2004), atomized, diffused.

Yet, since the 2000s, demand has been growing for another research direction, more focused on what people actually do with media and cities; how cities actually negotiate ICTs affordances; how media languages and sociotechnical systems remediate urban space. An important step has been Stephen Graham's research manifesto (2004) which criticised new media studies as "city-blind" and called for empirically-grounded research on the media/city nexus: a challenge that has since been accepted by many scholars. Yet this has proven to be an equally, if not more difficult undertaking than macro-analyses. In-depth work on cities always reveals a wealth of interconnected phenomena that require multi-disciplinary efforts to be properly addressed; focusing on communications makes no exception. Actually, it could be argued that most, if not all, contemporary urban processes feature a media component. As media become diffused, practices enacted by city users are increasingly intertwined and dependent on communication technologies. Thus media content becomes active in processes of socio-spatial production beyond the traditional (and well-studied) impact on legitimation and perception. This calls for media studies to rethink methods and theories to address this new object. This need is also shared by urban studies, since materiality, social actors, representations and practices are tied into a knot in which the elements can be addressed alone (or even in pairs), escaping the grip of critical theory to some extent.

When two complex objects like media and urban space come to interact, narrow analytical lenses must be discarded. In our view, this problem has haunted urban studies for some years now. By definition a cross-disciplinary field, Urban Studies has nonetheless shown a certain disdain to take up the analytical paradigms of media studies. In the 2010 world congress of RC21–a large urban studies international network–less than ten papers out of several hundreds were explicitly concerned with media. Yet if we are to understand how media and cities go together at the micro level we need to go beyond disciplinary borders: a by no means conclusive list of paradigms to mobilise would comprise semiotics, sociology, science and technology studies, urban studies, specialized media studies, audience studies, architecture and so on. We can perhaps summarise the core concerns of this line of research (which produced edited works including Eckart, 2008 and McQuire, Martin & Nederer, 2009) within three distinct areas: how people perceive and represent urban space; what they do with media in urban space; and what role is played by media in the social production of urban space itself. This tripartite structure also informs this volume, although each part shares concerns and perspectives with the other two.

Part one is dedicated to "Media and the Social Shaping of Urban Space"–that is, it examines the feedback processes between media and urban space. In Chapter I, Matteo Tarantino and Simone Tosoni illustrate a possible analytical model of spatial production processes and apply it to a conflict over a neighbourhood in Milan, Italy. While the conflict opposes migrants and Italian residents, the analysis allows to overcome simplistic readings of the conflict as an "ethnic" one (as a "culture clash") and brings to the fore the wealth of complex negotiations entailed by the conflict at the symbolic, physical and pragmatic levels. In Chapter II, Seija Ridell articulates in a McLuhanian framework the notion of the contemporary city as a "medium of media", able to re-mediate existing communication technologies and paradigms. Sami Kolamo's work (Chapter III) deals with how urban space is impacted by large-scale football events. Kolamo's analysis examines from a critical standpoint the negotiations of the needs of sport spectacle and media apparatuses and the specificities of the urban contexts in which such processes take place. In Chapter IV, Christian Oggolder mobilizes graph theory to uphold the argument of a loss of centrality of the city due to the increased density of ICT infrastructure. In Chapter V, Federica Timeto deals with a form of gamification of civic engagement, transformed in a series of missions through which users can gain points by improving city space (a topic which returns in Chapter IX). Finally, in Chapter VI, Moira Sweeney analyses how the representations of the Dublin docklands influence and reflect the actual renovation processes undergoing in the area.

Part two deals with "Media practices in urban context", or with how urban spaces interact with ICT-related practices enacted by subjects. Barbara Scifo's work (Chapter VII) draws from research on Italian youth to tell us how GIS-enabled camera-phones are used in the practices of Italian teenagers dealing with their position in space. Chapter VIII continues the reflection on mobile phones, as Satomi Sugiyama examines how the structure of Japanese public spaces and social norms constrains the practices of use of this technology in urban settings. In Chapter IX, Gabriele Ferri and Patrick Coppock discuss "urban games", or ludic practices enacted in and with urban space, and their potential for civic engagement. In Chapter X, Caruso, Fassone, Ferri and Salvador propose a taxonomy of ludic applications of geo-localisation in urban spaces: a framework that will be useful to future empirical studies of these practices. The concluding chapter of Part two (Chapter XI) deals with ICTs and urban catastrophe, as Emiliano Trerè and Manuela Farinosi examine how citizen journalism acted as a resource mobilized by the citizens of the Italian city of L'Aquila after a disastrous earthquake.

The third and final part of the volume deals with "City representations, media imageries and urban experience"–i.e. with the strategies through which social actors (individual or collective) represent urban space. In Chapter XII, Miriam De Rosa uses the case of an interactive film by Antoine Viviani composed by an assemblage of geo-localized "chapters" by different users, as an example of a renegotiation of urban imagery through database logics. Katalin Fehér (Chapter XIII) assumes as the lynchpin of her analysis the social construct of "city identity" and discusses how the "digital" part of this identity (i.e. the one emerging from the interaction of data flows) is produced. Gabriella Sandstig (Chapter XIV) focuses on a specific brand of urban perception: that of personal security. Sandstig's work draws from the perspective of cultivation theory applied to quantitative data to show how media representations of urban space security impact the mobility practices of the individuals. Chapter XV deals with communication and monumentality, as Gonca Noyan discusses how the symbolic value of an historical bridge in Mostar (Bosnia and Herzegovina) has been shaped by competing discourses throughout history. Walter Mattana and Marianna Ciancia (Chapter XVI) discuss *Imagine Milan*, a project for the collection of multimedia materials dealing with place identity in the capital of Lombardy. Their work shows the strong connection between place identity and the individual stories of inhabitants, thus stressing the connection between individual practices and shared representations of spaces. Finally, Amedeo D'Adamo (chapter XVII) specifically focuses on cinematic representations of urban space, discussing in particular the influence of Dantean architectural conceptions on the representation of cities in modern cinema.

References

Borja, J., & Castells, M. (1996). *The Rise of the Network Society*. London: Earthscan Publications.

Castells, M. (1997). *Local & Global: The Management of Cities in the Information Age*. London: Earthscan Publications.

Eckardt, F., Geelhaar, J., Colini, L., & Willis, K.S. (Eds.). (2008). *Mediacity. Situations, Practices and Encounters*. Berlin: Frank&Timme.

Graham, S. (2004). Beyond the 'Dazzling Light'. *New Media & Society*, 6(1), 16-25.

Graham, S. (Ed.) (2004). *The Cybercities Reader*. London: Routledge.

Humphreys, L. (2007). Mobile Social Networks and Social Practice: A Case Study of Dodgeball. *Journal of Computer Mediated Communication*. 13(1), article 17.

Humphreys, L. (2010). Mobile Social Networks and Urban Public Space. *New Media & Society*. 12(5) (August), 763-778.

McQuire, S. (2008). *The Media City*. Thousand Oaks, New Delhi, London: Sage.

McQuire S., Martin, M.A., & Niederer, S. (Eds.). (2009). *Urban screen reader*. Amsterdam: Institute of Network Cultures.

Mitchell, W. (1996). *City of Bits: Space, Place and the Infoban*. Cambridge: MIT Press.

Sassen, S. (1994). *Cities in a World Economy*. Thousand Oaks: Pine Forges Press.

Sutko, D.M., & De Souza e Silva, A. (2011a). Location-aware mobile media and urban sociability. *New Media & Society*, 13(5) (March), 807-823.

Sutko, D.M., & De Souza e Silva, A. (2011b). *Net locality. Why location matters in a networked world*. Malden, MA: Wiley-Blackwell.

PART I.

MEDIA AND THE SOCIAL SHAPING OF URBAN SPACE

CHAPTER I

MEDIA AND THE SOCIAL PRODUCTION OF URBAN SPACE: TOWARDS AN INTEGRATED APPROACH TO THE CONTROVERSIAL NATURE OF URBAN SPACE

MATTEO TARANTINO AND SIMONE TOSONI[1]

This chapter articulates an approach to sociospatial production drawing on social geography (first section), STS and media studies (second section), in order to address the complexity of urban space in contexts of pervasive ICTs. The third section will apply the model to an ongoing sociospatial controversy regarding the Paolo Sarpi area in Milan.

1. Space as a social product

Henri Lefebvre (1991) and his subsequent re-readings (e.g. Soja 1989, 1996; Mitchell 1996, 1998; Harvey, 1989) pushed the social sciences to overcome *a priori* conceptualizations of space in favour of its interpretation as a social product. Lefebvre's *trialectics* among *spatial practice*, *representations of space* (conceived, planned space) and *representational space* (space as lived and appropriated by social actors, also through imagination)[2], which can also be read from another

---

[1] We would like to thank Trevor Pinch for its precious commentary on the chapter.

[2] Unwin (2009, p. 18) remarks that many of Lefebvre's applications often simplify and reinterpret his concepts (see also Soja, 1996, p. 8). Interpretations of "representational space" are particularly numerous (Dimendberg, 1998). Our paper leans towards Harvey's reading of them as "mental inventions (codes, signs, 'spatial discourses', utopian plans, imaginary landscapes, and even material constructs such as symbolic spaces, particularly built environments, paintings,

perspective as the interplay of practices, perceptions and physicality (the social, mental and physical spaces of Lefebvre–1991, p. 11)[3], produces *human space*, "simultaneously material object or product, the medium of social relations, and the reproducer of material objects and social relations" (Gottdiener, 1985, p. 129). As Adams (2009, p. 175) remarks, with respect to other similarly tripartite distinctions (such as Sacks' nature/meaning/social relations, or Gould and White's environment/ behaviour/information), Lefebvre's scheme "covers the same terrain without using mutually exclusive terms" – thus maintaining "unity".

While not the first to stress the social *nature* of space (Unwin 2000, p. 12, points to the "long tradition" within geography "with the claim that space can be shaped from the social meanings", dating back at least to Kirk, 1952), Lefebvre's crucial contribution was to bring *imagined space* into the dynamic of spatial production as the cornerstone to a unified approach to spatial theory. Lefebvre's effort drew from Castoriadis' (1975) concept of "imaginary" and, arguably, Kevin Lynch's seminal work on the "image of the city" (Lynch, 1960) wherein the author illustrates the cognitive role played by "mental images" in the orienting practices of urban dwellers. "Imaginaries" is hereby intended as the ensemble of "representations" through which members of a social group "imagine their social existence" and that of their surrounding world and relationships (Taylor, 2004, p. 23). As such, they vary along cultural and ethnic axes (see for example Hayden, 1997; Arefi & Meyers, 2003): indeed, one of Lefebvre's main points is that "each society produces its own space" (1991, p. 29). For example, McCann (1999) uses Lefebvre's triad to understand the racial tensions underpinning a riot in Lexington by showing how imaginary urban geographies correlate strongly with representations of racial identities.

As a social product, space is a potential object of controversy among actors competing to establish a specific "space" as dominant. This is especially true for urban space, where governance models are plural and fragmented and transformation processes are constant. Such conflicts vary in scope. From his Marxist perspective, Lefebvre theorized a macro-conflict with the logic of capitalism trying to establish its own

museums and the like) that imagine new meanings or possibilities for spatial practices" (Harvey, 1989, p. 218).

[3] For a critique of this notion of "social production" see Unwin, 2009. For a distinction between "space" and "place" where social construction applies to the second but not the first, see Gieryn, 2000; also Harvey, 1996.

"representation of space" (vehiculated by the maps, projects and discourses of architects, planners and other specialists) as dominant: a functional, homogeneous, formalized "abstract space", to be superimposed upon the vital and differentiated space of "authentic" human life. This superimposition requires the erasure of the "history" of lived space, and of the very nature of this conflict. In this sense, being sites of authentic living (and not of formal abstraction), representational spaces open the potential for "thinking differently" about space, engaging in a dynamic of spatial resistance (1991, p. 39).

Particularly intense conflicts erupt over *public urban space,* the alleged "decline" of which represents a key preoccupation of Lefebvre himself and others (e.g. Lefebvre, 1991, 1996, p. XIV; Jacobs, 1968; Sorkin, 1992). Neal defines public space as "all areas that are open and accessible to all members of the public in a society, in principle though not necessarily in practice" (2010, pp. 1-2). Mitchell (1996, p. 2) identifies two main and competing definitions of public space (see also Mitchell, 1999, p. 128): as "a space marked by free interaction and the absence of coercion by powerful institutions [...], an unconstrained space within which political movements can organise and expand into wider arenas", and as a regulated retreat where a "properly behaved public might experience the spectacle of the city". Mitchell then applies Lefebvre to a controversy about the legitimate uses of a park in Berkeley, and argues that the first definition (unconstrained space) is akin to "representational spaces", whereas the second coincides with "representations of space".

Like Mitchell, McCann reads riots as practices of resistance against power-backed inscriptions of representations of space: i.e. riots are practices bridging representations of space and representational spaces, as they work "within the bounds of the conceived abstract spaces of planners and architects while simultaneously being shaped by individuals' perceptions and uses of space" (McCann, 1999, p. 151). Both McCann and Mitchell (along with others, e.g. Castells, 1983; Soja 2000; contrast with Pickvance, 1985) sympathise with the "unlawful" practices of their subjects and tend to consider them as legitimate acts of resistance, echoing Michel De Certeau's (1984) distinction between the "strategies" enacted by institutions and the "tactics" enacted by subjects. These authors tend to appreciate the "progressive" value of traditionally dysfunctional places such as "slums, *barrios* and *favellas* [sic] [...] as localised 'reappropriations' of space that may furnish examples of such 'representational spaces' or 'spaces of representations' by which certain sites are removed or severed from the governing spatialisation and returned to the realm of 'communitas'" (Shields, 1998, p. 165).

The Gramscian hegemony-resistance frame also informs studies regarding the role of media in processes of spatial production, which mostly assume mass media representations of actors and spaces involved in spatial conflicts as their foremost (and thus *isolated*) object. Through critical discourse analysis, these studies examine either how media representations strengthen institutional representations of space (e.g. Sundberg & Kaserman, 2007) or how local practices of discourse production and/or media usage enacted by marginalized communities (within mediated spaces of representations) act as means of resistance against institutional narratives. However, as suggested by Thrift (2004, p. 44) precisely in relation to De Certeau, the "humanistic romanticism" of this dichotomic reading of urban conflict (which ultimately sets physical space and architecture against immaterial resistance practices–see also Farìas, 2011) clouds an appreciation of the hybrid and intertwined nature of spatial practices–if we accept that "representational spaces", "spatial practices" and "representations of space" are into an actual relationship of co-construction (or, in Lefebvre's terms, co-*determination*).

The dualistic option is further weakened by the transformations brought about by (a) the increased presence of media in contemporary societies due to an increased portability of ICTs and (b) the "convergence" of media platforms towards a common digital codification, which has an impact on the temporal and spatial coordinates of practices of discourse circulation. This "ubiquity" produces continually evolving "mediascapes" (Appadurai, 1996) of artefacts, circuits and practices which, as Graham (2004, p. 4) remarks, were at one time studied as *substitutive* of urban space (e.g. Webber, 1968; Pascal, 1987), but are now increasingly understood in their interactive relationship with it. Indeed, mediascapes continuously supplement, extend, curb or otherwise negotiate all three of Lefebvre's "spaces"–and therefore profoundly impact the production of urban space.

We will return on these issues below: for now, let us simply state our starting point. We argue that the analysis of contemporary urban spatial conflicts requires an integrated approach that (a) maintains the dynamic nature of social space as a social product; that is, the notion that social space *exists* as the interrelation among physicality, representations and practices[4] (b) avoids the pitfalls of social and technological determinisms and (c) takes the role of the communication technologies into due consideration. We will outline such a model in the following sections.

2. Towards a unified approach

The socio-spatial approach frames urban space as a social product defining different sets of perceived affordances and constraints[5] for social action. This product undergoes continuous processes of co-shaping on three interrelated levels:

(a) Its *material* and *morphological* (i.e. pertaining to form and structure) dimensions, shaped for example by planning and construction processes.

(b) The *practices* it hosts, since each social practice contributes to shape and reshape the perceived affordances and constraints for any other social actor. Both the formal and informal rules of urban space usage participate to its social shaping, and the relationship among practices and space must be addressed as a relationship of co-construction.

(c) Its *social representations*, which influence the perceived character of affordances and constraints. These representations include descriptions of the morphology, definitions of legitimate practices, and depictions of all the social actors enacting these practices.

This approach to the social production of urban space involves two critical points:

(a) The vast array of social actors, with varying levels of power, who contribute to the shaping process at each level. The relationship of each actor's contribution with all the others' can range from mutual

---

[4] While we analytically separate "practices" and "representations", we are well aware that both perceptions and representations can be addressed not only in their content but also in their pragmatic aspect; that is, as "practices".

[5] While the original formulations of the concepts of "affordance" (Gibson 1977, 1979) and "constraint" refer to an objective quality of an object, irrespective of their perceptions or interpretations, subsequent readings (e.g. Norman 1987, 1990) switched the attention to "perceived" affordances and constraints, as actually structuring social actors' practices.

reinforcement to incompatibility and social conflict. Therefore, urban space can be addressed as a *controversial social object*.

(b) The intricate relationships among all the heterogeneous elements involved (morphology, practices, symbolic level, social actors), each of which shapes and is shaped by the others.

We argue that these points can be addressed by an approach borrowing concepts from both Science and Technology and Media studies.

2.1 Dealing with complexity I (with a little help from STS): Urban Space as a Controversial Object

The conceptualisation of urban space as socially produced enables a methodological dialogue with STS and their attempt to account for the processes of social shaping of technological artefacts, conceived as material objects and as sets of affordances and constraints for their users.

Attempts to bridge STS and Urban Studies have increased in number, with Actor-Network Theory (Callon, 1986, 1987, 1991; Latour, 1987, 1988, 1999) playing a lead role (culminating in Farìas & Bender, 2009; Farìas, 2011). For Aibar and Bijker (1997) this bridging has its origins in the mid-1980s along with a new attention towards "the role of politics and cultural norms and values in the shaping of urban technological systems. Urban technology is now put into the broader context of urban culture, politics, and socioeconomic activities (…). Technology is considered to be socially shaped, at least partially; it is no longer treated as a given, unyielding, and exogenous factor framing other dimensions of life in the city" (1997, pp. 5-6). Among these new STS approaches to the city (for a review, see Johnson-McGrath, 1997), Aibar and Bijker's worked on the Cerdà Plan for the Extension of Barcelona, conceiving town planning as a technology and the city as an artefact.

On those heels, Graham and Marvin attempted to integrate the leading approaches in STS (Social Construction of Technology, Large Technical System and ANT) with their studies on urban telecommunication infrastructure (1996) and on the phenomenon of splintering urbanism (2001). On the other hand Coutard and Guy (2007) criticised the "splintering" hypothesis, suggesting that thanks to a systematic dialogue with STS, Urban Studies could overcome their dystopian attitude and their "intellectually and politically disabling technological pessimism" (p. 713).

Gieryn (2002) borrowed from STS the theoretical tools of heterogeneous design, black boxing, and interpretative flexibility as "middle range" concepts to make "the abstractions of 'structuration' and 'reproduction' (…) more friendly for empirical analysis" (p. 45) and used them on the social construction of buildings (a laboratory at Cornell University), conceived as an attempt to stabilise the connection between agency and structure. Hommels (2005; see also 2005b), addressed urban obduracy through SCOT, ANT and LTS, clarifying the implications, focuses and potentialities of each approach, while also renewing the call to a dialogue with STS, judging the work done so far episodic and discontinuous at best.

The works we mentioned share a common focus on the material production of urban space (the first point of our model) or on the shaping of urban technological infrastructures. However, this narrow focus appears to be more related to STS's traditional disciplinary interests than to a true theoretical option. We argue that STS can provide powerful analytical tools to approach urban space as a multi-layered (physical, pragmatic and symbolic) social artefact. Our attempt will rely on an eclectic borrowing and re-adaptation of theoretical and sensitising concepts (Blumer, 1954) derived from different approaches, and in particular from ANT and SCOT.

Symbolic representations of space have a pivotal role for us; we define space as always potentially controversial; and we acknowledge the plurality of relevant social actors that concur to the social production of space. These three options echo the framework of the SCOT approach, originally proposed by Pinch and Bijker (Pinch & Bijker, 1984; Bijker, 1987, 1995; Pinch, 1996, 2003, 2009; Pinch & Kline, 1996). Bearing in mind that SCOT constitutes more of a flexible interpretative strategy than a formulaic and standardized methodology, the approach can be summarized in four main points. "First, the notion of a relevant social group is introduced. Such a group is defined as a group which shares a particular meaning of the technology. (…) The second part of SCOT is the idea of 'interpretative flexibility' which (…) is a notion developed in the study of science. This idea points to the radically different meanings which technologies can acquire for different social groups (…). The third key element of SCOT is the process of closure or stabilisation whereby the interpretative flexibility of an artifact vanishes. Particular closure mechanisms can be identified which lead to some meanings vanishing" (Pinch, 1998, pp. 9-10). The last point relates "the content of a technological artifact to the wider socio-political milieu. (…) Obviously, the sociocultural and political situation of a social group shapes its norms and values, which in turn influence the meaning given to an artifact" (Pinch & Bijker, 1984, p. 46). When adapted to the social construction of space, this approach provides a threefold focus to empirical research:

1) It drives the researcher to pay a peculiar attention to the different social actors involved in the social shaping of urban space. In our model, this involvement potentially pertains to all three levels: the actual planning, construction and modification (institutional or otherwise) of urban space; the shaping of its set of perceived affordances and constraints through actual practices (and/or through the political definition of its legitimate uses), and through its explicit or implicit representations. The analytical prominence assigned to social actors and their groups must not be mistaken for social determinism. In fact we discard any deterministic relationship of causation between socio-cultural coordinates, spatial practices and representations in favour of a case-by-case empirical analysis. This process of social shaping feeds back on the social actors themselves, contributing to their definitions (on this, see also Tosoni & Tarantino 2013).

2) SCOT's concepts of "interpretative flexibility" and "closure" call the researcher to account for the plurality of potentially conflictive meanings assigned to artefacts by relevant social actors, and to the stabilisation of these meanings. On this basis, with "controversial nature of urban space" we refer to the negotiations among relevant social actors, and with "institutionalisation of space" to their temporary stabilizations. Such negotiations can be direct or mediated (e.g. through material artefacts, juridical systems, or ICTs), and they pertain to all levels of our model (thus not only the representational level). The operations performed by social actors on the morphology of urban space, the uses of space enacted and the practices performed, and the symbolic meanings assigned can in fact be synergic or conflictive. As the stabilisation of these shaping processes is always temporary and reversible, we prefer to use "institutionalisation of space" and "obduracy" rather than "closure" to describe it. The "interpretative flexibility" of urban space never vanishes, and closure can rarely be achieved at any of the three levels: the institutionalisation and obduracy of urban space is the result of a reiterated operation of stabilisation of practices and representations, even when morphological changes appears to have reached a "closure". Indeed, obduracy and stabilisation call for a sociological account as much as conflict and change, although spatial controversies and conflict offer vantage points from where to observe processes of social shaping

(Venturini, 2008) insofar as they intensify the discursive production and conflictive practices of social actors.

3) The last point of SCOT addresses the micro-macro link. On this respect, STS provides analytical concepts (Akrich & Latour, 1992) traditionally associated with ANT potentially useful to jointly address the "broader context", the complex relationships among all the levels of our model, and the co-shaping of actors and space. ANT's concept of "network" reframes the otherwise unsolvable conundrum of the macro-micro link. The network does not feature different levels to be linked analytically: neither micro nor macro, neither local nor global, its extent is defined only by the ability of its components to influence each other: "even a longer network remains local at all points" (Latour, 1993). Laws clarifies this methodological stance: "if we want to understand the mechanics of power and organisation it is important not to start out assuming whatever we wish to explain. For instance, it is a good idea not to take it for granted that there is a macrosocial system on the one hand, and bits and pieces of derivative microsocial detail on the other. If we do this we close off most of the interesting questions about the origins of power and organisation. Instead, we should start with a clean slate. For instance, we might start with interaction and assume that interaction is all that there is. Then we might ask how some kinds of interactions more or less succeed in stabilising and reproducing themselves: how it is that they overcome resistance and seem to become 'macrosocial'" (1992, p. 2).

This means deconstructing the macro-micro link by assuming an in-between analytical position focusing on the interactions of all "actants" (as ANT calls the elements of the network, human or not) that *participate* to the network—that is, that interact and are co-shaped by this interaction. Such a stance does not imply narrowing the focus to micro analysis: as Law remarks, "arguably it is possible to impute somewhat general strategies of translation to networks, strategies which, like Foucauldian discourses, ramify through and reproduce themselves in a range of network instances or locations (…) What might such strategies look like? This, again, is an empirical matter. But since no ordering is ever complete, we might expect a series of strategies to coexist and interact" (1992, p. 7). Simply, any social process being analysed, whatever its scope, must be addressed in its specific "localized" forms, that is the material modification of space it requires, the practices it enables, fosters or blocks in a specific place, and the symbolic representation it implies.

Moreover, by emphasising the relationships among the connected elements over their nature, the concept of "network" is compatible with the heterogeneous nature of urban space as theorized by our model. Farìas acknowledged the importance of such an approach for Urban Studies: "Clearly the notions of space as produced by means of different practices, strategies and circulations, and of space as multiple, are important shared starting points facilitating the dialogue between Urban Studies and ANT" (Farìas, 2011, p. 22; authors' translation).

Heterogeneous elements (material, pragmatic and symbolic) can be connected by a relationship of mutual co-shaping. ANT's concepts of "heterogeneous engineering" (Law, 1987) and "heterogeneous design" (Gieryn, 2002) describe these kinds of connections during the engineering and planning phases respectively. Instead, we prefer to borrow the concept of "translation" (Latour, 1992; Law, 1992) to account for the fact that the social shaping of urban space is always ongoing. The concept was used by Adams to account for the relationship among ICTs and space: "Are we interested in what a particular text or discourse has to 'say' about a particular place, or in how position is affected by connections and flows, or in how interconnections constitute a particular social space of inclusion and exclusion, or in how places accommodate media and media become ingredients in places? If none of these alternatives make sense, we may be looking at communications associated with translations, hybridity, actor networks, horizontality, and a 'flat ontology'. In this case, communications are no longer seen as arcs between here and there or this and that; they instead become integral to the here-ness and there-ness, this-ness and that-ness of boundless selves" (Adams, 2010, p. 50).

The main aim of our approach is thus to analyse the various ways in which the symbolic, the pragmatic and the material translate into one another.

Social actors are in no way excluded from these relationships of translation: the relationship among actors and the material, pragmatic and symbolic nature of space is always a relationship of co-shaping. On this respect, while ANT's principle of radical symmetry between human and non-human holds significant theoretical value, from a methodological perspective we think that SCOT's attention to social actors and to their controversial relationships with their contentious object (in our case, space) represents a useful starting point to unfold the complexity of the heterogeneous network that constitutes urban space, and a crucial orientating principle to render this complexity empirically manageable (Pinch, 2009, 2010). Without a measure of relevance, the rhizomatic complexity of the infinite interconnections of the network would be untranslatable into academic discourse. We think that this measure of relevance is co-constructed by the interplay among the social actors' controversial relationship with space and the analytical traditions we have advocated to tackle urban space.

2.2 Dealing with complexity II (with a little help from Media Studies): the relationship between representations and social actors

Addressing "social space" as the interrelation of physicality, practices and representations through the combined recourse to sociospatial and sociotechnical approaches must be backed up by communication studies, since media play a key role at all the three levels. As Jansson remarks, "the linkage between geography and communication lies in the fact that (a) all forms of representation occur in space and (b) all spaces are produced through representation" (2005, p. 1). Bridges between media studies and sociospatial theories are in fact growing in numbers: media studies are slowly taking a "spatial turn" (Falkheimer & Jansson, 2006), and geographic studies a "communicational turn" (Adams, 2009, p. 10). In our approach, to account for the role played by media in processes of spatial production means to observe how communication technologies interact with the sociospatial triad–on all three levels and *at the same time*:

a) ICTs change the *morphology* of the space in which they are deployed. Billboards, cameras, screens impact the cityscape and therefore the perception of space by social actors (thus influencing the other two levels: their practices and representations). ICTs play a key role also at a macro level, supporting processes of transformation that impact on the morphological transformations of specific spaces: e.g. by influencing the centre-periphery relationship (see e.g. Fathy, 1991; Schuler, 1992), by attracting capital and expertise within urban centres, and by interconnecting the city to the transnational space of flows (Castells, 2004; Sassen, 2000).

b) At the level of *practices*, ICTs have an impact on the perceived affordances and constraints negotiated by social actors with social space– as well as on behavioural codes and coordination practices. CCTV cameras can change patterns of use of public space; mobile phones can reshape behavioural conducts in public space (Fortunati, 2002; Katz & Aakhus, 2004; Tosoni, 2009; Humphreys, 2005, 2010).

c) At the level of *representations* (i.e. projected by actors involved in sociospatial production on each other and on space) the role played by ICTs is twofold. On the one hand, mediated social representations exert influence on city images and imaginaries (Çınar & Bender, 2007a) in respect to such things as desirability, opportunities for job and leisure, and security (Vale, 1995)–thus impacting the other two levels of sociospatial production (e.g. in terms of housing and/or mobility patterns). On the other hand, ICTs offer potential discursive spaces to social actors. In particular, sociospatial controversies trigger actors to produce representations of the self (self definitions, as lynchpins of actors' practices of identity construction), of other stakeholders and of (contended) space.

It is important to stress how actors produce these representations in dialogue within pre-existing and possibly competitive representations (e.g. during spatial controversies; see Tosoni & Tarantino, 2013): the ensembles of representations that amount to SCOT's "interpretative flexibility" tend to be in a relationship of dynamic interdependency. This dialogue can be (and mostly is, in advanced societies) technologically mediated; however, we argue that notions of media as a potential common public arena or public sphere are too simplistic. Elsewhere (Tosoni & Tarantino, 2013) we discussed the concept of *media territories* to describe how social actors segment the overall ensemble of communication technologies and platforms available in the convergent media system (De Sola Pool, 1983; Jenkins, 2006) around a specific issue. Social actors do not communicate through a "channel" but through a temporary assemblage of media that mediates their representational dialogue, and such "territories" can be *contiguous* (when the different actors share access to platforms and/or channels), or partially not contiguous (for example, when actors do not possess the necessary skills or will to access a platform/channel). Of course, actors may include in their media territories platforms or channels which they infer as being part of other actors' territories, with the intent of monitoring them and strategically adjust their own conflictive representation accordingly (including self-representations). The relevance of media territories problematizes the thesis of a generic "media empowerment" granted by the convergent media system to all actors involved in controversies. While we don't exclude that the increased accessibility of digital media may grant a new reach capability (Castells,

1997, 2007; Atton, 2004; Russell, 2005) and the ability of circumventing the gate-keeping of "traditional" media (Gitlin, 2003), these cannot be assumed without a careful scrutiny of the media territories of all the social actors involved in a controversy. When dealing with the *actual* spaces of representations which participate to sociospatial production, the complexity of these dynamic interrelations must be taken into account: in Tosoni & Tarantino 2013 we proposed a detailed methodological approach, drawing on Hine (2000), to map media territories and their dynamic interrelations.

Increases in ICTs convergence and ubiquity further intertwine the three levels of spatial production–especially so during sociospatial conflicts, when discursive production is enhanced, media territories see their geographies change, and practices of spatial appropriation are continuously renegotiated. This strengthens the need for interdisciplinary, integrated paradigms to conflictive sociospatial production, drawing on STS urban and communication studies (see Figure 1). We will apply our proposal to a case study in the third part of this chapter.

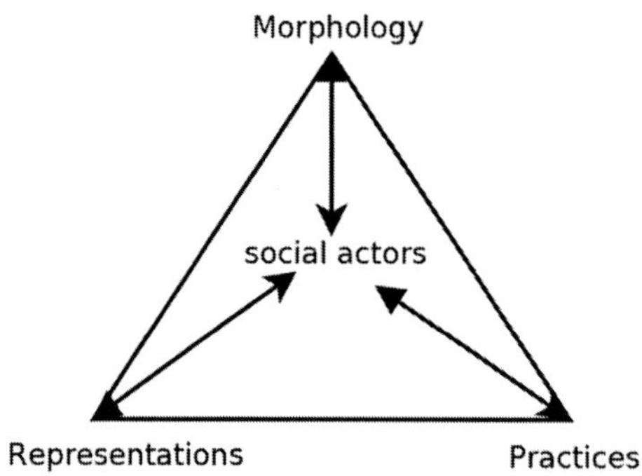

Figure 1 - The relationships of co-shaping that altogether make up "social space".

3. A case study: The Conflict for Paolo Sarpi

We will illustrate our model through a case study regarding an open controversy over the Paolo Sarpi area in Milan (see also Tarantino & Tosoni, 2009a, 2009b, 2011, 2013; Gelpi & Tarantino, 2008). The controversy is multi-dimensional and very complex and so far, in SCOT's terms, it has reached no "closure". As such, our work is also still ongoing:

this paper will deal with the 2007-2012 period and refer to previous works when necessary.[6]

"Paolo Sarpi" (named after its main road) is an area in central Milan delimited by the streets via Canonica, via Elvezia and via Bramante. Albeit mostly inhabited by Italians (Manzo, 2009) in public discourse the area is known as the "Milanese Chinatown" and constitutes the symbolic centre of the Chinese presence in the city, because of its massive concentration of Chinese businesses and services. The area is characterized by narrow streets (with a width of 7 to 5 meters or less). Diachronic analysis of city maps shows a largely unchanged structure, although, as we will see, the uses of its space have changed considerably.

The controversy is mostly related to the relationship between migrant usages of the area and its morphology, and has gained visibility since the 1990s, peaking in 2007 when it triggered an incident which is considered as the first Italian "ethnic riot". The 1990s marked a change in the Chinese practices of space usage: from semi-clandestine leather shops, invisible to the public, migrant businesses shifted to wholesale clothing retail. This change in practices changed the morphology of space: Paolo Sarpi public space became encumbered by trucks, trolleys and other means of transportation for goods. The success of these businesses attracted other migrants to the area, and a diversified range of Chinese-operated activities flourished–including travel agencies, restaurants, libraries, video rental stores. The new practices of use needed signs and insignia in Chinese; therefore the visibility of the migrant presence increased.

Thus the spatial controversy became manifest. Starting from the late 1990s, a part of Italian residents gathered in associations, with the Vivisarpi association assuming a prominent role, and started lobbying for a curtailing of the wholesale businesses. In 2007 the municipality responded to this lobbying by issuing a strict repressive system against docking-undocking activities taking place outside of the parking areas and

[6] Our empirical effort entailed: (a) Ethnography in the area; (b) In-depth interviews with Chinese (n=25) and Italian informants (n=8). (c) Online ethnography of pertinent websites, BBS, blogs and forums, both Italian and Chinese. (d) Content analysis of Italian media coverage, using as sources the newspapers *La Repubblica* (n=125 articles), *Corriere della Sera* (n=99), *Il Giornale* (n=5), *Libero* (n=5), *Europe China News* (n=17) and the *Rénmín Rìbào* (n=1); 47 Xinhua press releases; 10 Italian and 2 Chinese TV newscasts (e) Analysis of Milan city maps throughout the years; (f) Analysis of the real estate market using 2012 data from the Milan Chamber of Commerce.

allowed hours. This system seriously hindered otherwise legitimate Chinese businesses. Discontent among migrants increased, until in April 2007 a riot pitted an estimated 300 migrants against the city police (for a description of the riot and its causes, see Tarantino & Tosoni, 2011). Subsequently, several projects for relocating Chinese wholesale activities were discussed; all of them failed. In 2009 Paolo Sarpi Road was made a limited traffic zone (LTZ) monitored by CCTV, allowing business-related traffic only during specific time windows. However, Chinese wholesale activities have remained dominant in the area, as they mostly operate outside of the main road.

We will address this sociospatial controversy on the three analytical levels of physicality, representations and practices. As these levels are in a relationship of co-shaping, any one would represent a good starting point: our option for SCOT drives us to start from a map of relevant social actors and their representations of public space (Tarantino & Tosoni, 2009a, 2009b). In public discourse, the main actors involved in the controversy appear to be Italian residents of the area, the Chinese "community" and the Milan municipality. However, closer scrutiny starting from shared representations reveals a much more diversified galaxy of actors. First, Italian residents can be divided in at least two groups: those aligned with Vivisarpi and those who are not, including some of our informants. Second, although often represented as such, Milanese Chinese migrants cannot be considered a "community" (Tarantino & Tosoni, 2011), being much more dispersed and stratified (by generation, class, place of origin). This has hindered the migrant's ability to produce recognizable leaders. Moreover, as we will see, Chinese retailers and wholesalers are not always aligned.

The media territories articulated by actors around the controversy, mostly appeared to be *not* contiguous, also because most migrants and Italian residents did not possess enough skills and knowledge (e.g. linguistic) to access the other actors' media systems. Thus, for example, a series of bilingual Italian-produced webcasts about life in Paolo Sarpi was largely ignored by migrants (Gelpi & Tarantino, 2008). Conversely, although Chinese television and newspapers covered the incident, no sign of this content made its way into the Italian media (Tarantino & Tosoni, 2011, 2013). Chinese migrants mostly watched Chinese satellite TV, accessed Chinese-language Internet and read Chinese newspapers (especially local ones). Some second-generation migrants constituted an exception–being proficient in Italian–so a kind of "weak" contiguity could be reached in some cases. No sign of a comparable phenomenon (subjects proficient in Chinese acting as mediators and agents of contiguity) was registered with respect to Italian media territories.

The riot radically, if only momentarily, reconfigured the spatial territories of Chinese migrants–and it did so through spatial practices. As said above, riots can be read as radical practices of spatial re-appropriation transforming public space into a means of communication (bridging spaces of representations and representational spaces); and indeed, through the riot, Chinese migrants successfully created a contiguity with Italian media territories, to make their dissatisfaction known. This practice of using space as a medium continued in the days immediately after the riot, as the Chinese plastered the area with photographs of violence during the riot (taken either from the Internet or from local Chinese-language newspapers) and exposed in their shop windows signs in Italian lamenting the situation (Tarantino & Tosoni, 2009b). This attempt to bypass the mentioned limitations in platform access lasted just a few days; subsequently, signs and pictures disappeared from public space–only to be fed back into the convergent media system as pictures-of-pictures and pictures-of-signs which have since then been circulating in the Internet.

Through the analysis of these territories (Tosoni & Tarantino, 2013) three competing "representations of space" emerged as able to coagulate actors beyond their mentioned internal divisions. We will use these representations as starting points to illustrate, through their connected practices and morphological changes, the multiple processes of spatial production which are active in the area.

The first representation, a key resource for the Italian residents aligned with Vivisarpi, identifies the "true nature" of Paolo Sarpi as that of a *residential area.* In this representation, the area's position, public transport connections and the good quality of real estate fail to guarantee a high QOL because of the systematic diversion from its natural vocation due to the illegitimate usages enacted by a specific social group–the Chinese–who are represented as having suddenly invaded the area, starting to use it as a "working space" (see later). This representation relies around the narrative of a mythological past and of its destruction, with the old (and "true") Paolo Sarpi described as a sort of self-sufficient community village, providing for the necessities of daily life through a rich variety of shops and services (from barbershops to bars, from groceries to newsstands) - "third places" (Oldenburg, 1989, 2000) which also fostered a social life in public spaces described as rich and warm.

Indeed, the destruction of this "ideal way of life" is imputed first to the destruction of the *public* character of space by the spatial practices connected to wholesale activities, considered as incompatible with any other usage of space and rendering it *un-walkable* (being encumbered by illegitimate practices)–and, as such, un-liveable.[7] Establishing this representation of space as dominant is a cornerstone of Vivisarpi's action. Vivisarpi's media territory features the Internet as its foremost platform, and visual representation as its main code. Discourse here tends to over-emphasise the opposition between "what is" (shown) and "what could be" (remembered and narrated) (i.e. between spatial perception and representational space) in order to gain local consensus and legitimate authoritarian interventions.

As this representation legitimises specific instances of spatial transformation, Vivisarpi and the municipality attempt to translate it into spatial morphology through a heterogeneous strategy mobilising "intermediate translations". This strategy relies upon legal measures curtailing the "disruptive" processes, i.e. illegitimate uses of space, substitution of "third places" with "private" business, and the inscription in space of signs of "otherness" such as Chinese shop signs. Being perfectly legal, the wholesale businesses could not be simply shut down. To drive them out, the strategy pushed for a normative reshaping of the affordances and constraints of public space through a re-interpretation, extension and rigorous application of the existing norms. Throughout the years this strategy has forced all involved social actors to change their practices.

First, municipal authorities were enroled to foster the systematic and rigorous application of parking rules through the city police. These rules

[7] Notably, the depiction of the present-day Paolo Sarpi differs from those typical of other social groups hostile to immigration (Tosoni, 2007) insofar as space is not depicted as *unsafe* but as *encumbered* and *deteriorated* by its illegitimate uses. Yet in the early 2010s the "unsafety" began to emerge in Vivisarpi's representations. Its president was quoted in 2011 suggesting that the 'ndrangheta could be "behind Chinese wholesale activities", and that the area was "hostage of Chinese youth gangs" blackmailing local shops – while at the same time acknowledging that they are "invisible" and "this is what scares us the most". See www.ilgiorno.it/milano/cronaca/2011/10/17/601746-paolo_sarpi.shtml.

fit perfectly with the morphology of the area (narrow streets, lack of parking spaces) in trying to block Chinese activities through systematic fining, which damages their earnings. To this end, human (a permanent police garrison) and technological means (a CCTV system aimed at monitoring the space) were employed since 2007.

Migrants responded by changing their spatial practices, substituting trucks and cars with another technology–the cartwheel. Suppliers and customers could now park outside the area (where the same parking rules apply, but not the same human and technological means for their enforcement) and carry their goods to and from their businesses. Once again Vivisarpi lobbied for a re-interpretation of existing norms on the usage of sidewalks in order to frame cartwheel transportation as an illicit spatial practice. While the lobbying was successful, the use cartweheels effectively de-powered the technological means for the monitoring and fining of illicit spatial practices–as cartwheels lack plates or other clear identification features.

The legal system was mobilized not only to regulate spatial practices but also to reshape the morphology of the area. The application of old (and otherwise abandoned) norms forbidding the exclusive use of foreign languages for public signs (including shop signs) were advocated as means to block the "visual alienation" of the area. As mentioned, an even more radical morphological reshaping declared Paolo Sarpi Road a Limited Traffic Zone, effectively forbidding non-resident traffic.

Moreover, Vivisarpi-aligned social actors attempted to enhance the obduracy of the new legislative status through morphological changes, by inscribing the "ideal community" representation in space. Urban furniture–benches, car bumps, trees and flowerbeds–appeared in the area, paid by local retailers. Bars and restaurants were invited to "occupy" sidewalks with outdoor tables and tents. This strategy was only partially effective in blocking cartwheeling, as it only regarded the main road and did not affect the rest of the area, where the migrants' "illegitimate" uses of space continued, thus increasing the area's fragmentation.

The vigourous representative effort backing the strategy involved both physical and media spaces. In public space, social actors organised events such as fairs about "traditional Italian products", which were all translated in the mediascape and especially on the Internet. Again, space is employed as a medium and the events as symbols, and they are fed back into the media system, where they gain in permanence and reach (within specific media territories). Notably, particular Chinese events (such as the New Year celebrations) are not opposed by Vivisarpi, whose support of "Italian-ness" is presented as a mere attempt to counterbalance the trend towards ethnicisation of the area.

Concerning media, communicative campaigns have been enacted to promote local Italian businesses. Those include the SarpiDOC campaign, aimed at promoting the area as a business improvement district.[8] As we will see when discussing the third representation, this communicative strategy seeks a temporary mediation with the potentially conflictive interests, representations and strategies of retailers and leisure operators.

Figure 2 - A representation of the socio-spatial transformation towards the "ideal village", fed back into the Internet. Text: "Sarpi Road" "December 2009" "January 2011", Source www.melinsarda.org.

[8] Notice how DOC, or "controlled designation of origin", is a quality seal used for food and wines, and is synonymous in common talk with "authentic"

The second representation presents Paolo Sarpi as a *space for work*, and is a key resource for Chinese businessmen operating in the area. Our ethnographic work in the area and in the related media territories enabled us to identify another lynchpin of the migrants' strategies of identity construction, much more effective in terms of social cohesion than the (as mentioned, fragmented and problematic) notion of Chinese-ness: that of the migrants as *merchants* (see also Tarantino & Tosoni, 2008b; Gelpi & Tarantino, 2009). Migrants tended to measure the success of their migratory project–their "integration"–not in terms of positive sociocultural relationships with the hosting context, but mostly in terms of economic success (it must be stressed again that Chinese migrants by and large do not reside in the area, preferring to commute from cheaper zones). The social space of the merchant does not need to be ethnically or culturally connotated (if not for economic reasons, e.g. Chinese insignia to attract Chinese customers): it just needs to be functional for practices that are economic (and not cultural or political) in nature. However, this makes Paolo Sarpi no less symbolically invested for merchants–especially since Chinese merchants legitimately purchased it, a heavy investment which often implied substantial borrowings from friends and family. This representation of space has always been dominant for migrants operating in Paolo Sarpi. However, until the 1990s, the invisibility of the migrants' activities made them compatibile with the context insofar as they mostly shaped *private* and not *public* space. In 1998, limitations to migrants' enterprises requiring a bilateral agreement with the homeland were lifted, allowing Chinese businesses to emerge and subsequently switch to wholesale. This change brought along new spatial practices which negotiated the morphological constraints of the area by occupying sidewalks and lanes; each subsequent normative constraint was negotiated through the adoption of different transportation technology (from trucks to vans, private cars, cartwheels and bicycles).

Also this spatial representation rests upon a mythic past (Tarantino & Tosoni, 2011) where Paolo Sarpi used to be a run-down, unsafe neighbourhood, populated by other migrants (gypsies were often mentioned in informants' accounts). Chinese ingenuity, restlessness and industriousness are described as having revitalized the area (incidentally, driving away all the "really dangerous" migrants) through economic success, boosting real estate value. Therefore, the hostility of Italian residents is described as betrayal and ingratitude, and, being otherwise unexplainable, as motivated by *racism*. This representational shift actually ethnicised the conflict: even though the riot was effectively much more of a "merchant riot" than a Chinese one, some of the practices enacted during the episode (such as the waving of Chinese flags) signalled a temporary mobilization of the ethnic identity for cohesion purposes (which was confirmed and saluted by our senior informants as very positive, if ephemeral) (Tarantino & Tosoni, 2009a). However, this further ratified and radicalized the negative representations of Chinese migrants as "conquering aliens" (see Carini & Tarantino, 2010), possibly worsening the general social representation of Chinese migrants.

On a larger scale, this representation frames Paolo Sarpi as a hub for the transnational flow of goods mostly imported from China and distributed throughout Italy, thus altering the positioning of the city within the "space of flows". Moreover, as a hub, Paolo Sarpi actually *benefits* from a spatial concentration of wholesale activities: the logistics is much more sensible for both customers and suppliers, enhancing the strategic value of the area. In other words, this spatial representation renders the set of practices enacted by Chinese wholesalers not only legitimate but also beneficial and desirable. Therefore, Chinese merchants (along with Chinese service providers and retailers who flocked to the area after its expansion) have continued to "secrete" (in Lefebvre's terms) their social space through their spatial practices and daily routines. This secretion of space in turn strengthens in Vivisarpi-aligned actors a competing representation of the "ideal community" and its practices and representations, which, allied with political power, produce a different space co-existing with this one.

To further complicate the matter, a third representation, fostered by retailers and leisure operators (both Chinese and Italian) but also by a part of residents not aligned with Vivisarpi, has started to emerge: that of Paolo Sarpi as a leisure area for shopping, cultural consumption, nightlife and entertainment. Here the "Chineseness" of Paolo Sarpi would be a resource and a competitive advantage for a successful positioning within the network of Milanese leisure zones. This spatial representation features specificities that destabilise existing equilibriums and allegiances. On the one hand, the massive presence of Chinese wholesale businesses fragments and dilutes the presence of leisure businesses, effectively slowing the desired process of functional specialization. For this reason, social actors sharing this representation have often sided with Vivisarpi, supporting 2007's "parking crackdown", subsequent attempts to relocate wholesalers out of the area, and the related representational efforts (like the "natural shopping mall"). They were also particularly active in the beautification campaign (to the point of being involved in the planning process) that followed the institution of the LTZ, and contributed to the inscription of the normative constraints in the morphology of the area. On the other hand, the representation of Paolo Sarpi as a leisure zone is incompatible with the "ideal community" under many respects. The institution of the LTZ itself, and the idea of urban space it implies, is a first and central point of controversy. "Leisure" social actors generally supported the lobbying activities that lead to the LTZ, yet its reception has not been unanimously positive. Milanese retailers have traditionally been hostile to pedestrializations, considered as hindrances to good business; coherently, they strongly opposed any further extension of the LTZ beyond Paolo Sarpi Road. Leisure operators–especially restaurant owners–share these worries: an extension of the LTZ would make the area difficult to reach, thus less competitive in respect to other comparable areas of Milan. As a general stance, these social actors consider the process of morphological reshaping that led to the LTZ as (satisfactorily) closed, and oppose any further extension. Those who do not share this stance appear to prefer a complete abolition of the LTZ over its extension. The second controversial point relates to the embedding of signs of Chinese-ness into spatial morphology, perceived as potentially catering to two market segments: Chinese migrants and tourists. For the first, Paolo Sarpi possesses great symbolic value as "the place to go" for Chinese-related activities, businesses and services. With respect to the second, Chinese-ness would be functional to an "ethno-chic" characterization of the area, capitalising on the radical "otherness" of Chinese migrants (including the illegibility of their spatial inscriptions) to attract subjects looking for experiential tourism (Coen, 2004), analogously with what happened with San Francisco's Chinatown (Salter, 1984; contrast with Christiansen, 2003). Indeed, these social actors consider Chinese spatial inscriptions so ripe with potential that (unsuccessful) attempts have been made for their extension (e.g. "traditional" Chinese dragon gates at each end of the main road). These attempts have been fiercely opposed by Vivisarpi-aligned residents, and were also met with substantial disinterest by merchants.

These three coexisting and largely incompatible processes of spatial production altogether render Paolo Sarpi's social space fractured, discontinuous and contradictory. Their analysis enables a clearer view into the reasons behind the instability of the area. The social actors involved are plural and fragmented, and their allegiances appear to be mostly tactical. Moreover, the peculiarities of their media territories make an effective spatial dialogue difficult. Indeed, tensions in the area have remained high, although no further episodes of violence have been recorded since the riot. The sociospatial controversy has not reached any closure, although Vivisarpi's process has been to some extent successful with respect to the main road, especially after the morphological inscriptions. Yet this configuration appears only temporary, especially as a new, large-scale project of spatial transformation is currently aiming at connecting the neighbourhood with the recently revamped high-value area of Garibaldi. This project is changing the power balance of the involved social actors: more fieldwork will be needed to assess its impact on the currently schizophrenic space of Paolo Sarpi.

Conclusions

The application of an integrated model of sociospatial production to the Paolo Sarpi controversy allowed to overcome simplistic analyses of the conflict (such as the "ethnic" thesis) and to identify the unstable spatial identity of the area as the main issue. This instability is due to the co-presence of at least three incompatible processes of spatial production, each featuring specific representations and practices. Their interaction is further complicated by the non-contiguity of the media territories of involved social actors, which renders spatial representations essentially impermeable and, therefore, prone to radicalization and conflict.

The recourse to STS enriched the understanding of spatial production dynamics, reframing some of Lefebvre's formulations while preserving the key notion that the three levels of physicality, representation and practices move together at all times, and must be analysed as such.

Finally, the Paolo Sarpi case provides some evidence to dismiss many "humanistic romantic" readings of spatial conflicts–for it is precisely the "weak" subjects here (Chinese migrants) who try to inscribe Lefebvre's "abstract space" of capitalistic logic over "lived space", paradoxically defended by the "strong" subjects (Italian residents, backed by the municipality). When migrants enact resistance practices to the dominant "ideal community" representation of space progressively (and forcefully) inscribed in the morphology of the area (in itself, of course, a different inscription of capitalism), they do so to preserve wholesale retail businesses–which are not exactly Lefebvre's idea of "right to the city" but, as we said, are a fundamental part of their self-representation–for all intents and purposes, of their *identity*. In our view, this challenges essentialist notions (and narratives) of "human space" as opposed to "rationalized space", and highlights the need to account for multiple "humanization" and "rationalization" instances active in social processes– including the social production of urban space.

References

Adams, P. C. (2009). *Geographies of Media and Communication*. New York: Wiley-Blackwell.

—. (2010). A taxonomy for communication geography in *Progresses in Human Geography*, 35(1), 37–57.

Akrich, M., & Latour, B. (1992). A Summary of a Convenient Vocabulary for the Semiotics of Human and Nonhuman Assemblies. In W.E. Bijker & J. Law (Eds.), *Shaping Technology/Building Society: Studies in Sociotechnical Change* (pp. 259-264). Cambridge, MA: MIT Press.

Aibar, E., & Bijker, W. E. (1997). Constructing a City: The Cerdà Plan for the Extension of Barcelona. *Science, Technology, & Human Values*, 22(1), 3-30.

Atton, C. (2004). *An Alternative Internet*. Edinburgh: Edinburgh University Press.

Barra, L., Tarantino, M., & Tosoni, S. (2011). Etnografia Virtuale e Convergenza: spazi, tempi discorsi e pratiche della TV sul Web. In S. Tosoni (Ed.), *Nuovi Media e Ricerca Empirica*. Milano: Vita e Pensiero.

Bijker, W. E. (1987). The social construction of Bakelite: Toward a theory of invention. In W. E. Bijker, T. P. Hughes & T. J. Pinch (Eds.), *The social construction of technological systems: New directions in the sociology and history of technology* (pp. 159-187). Cambridge, MA: MIT Press.

—. (1995). *Of Bicycles, Bakelites And Bulbs: Toward A Theory Of Sociotechnical Change*. Cambridge, MA: MIT Press.

Blumer, H. (1954). What is wrong with social theory? *American Sociological Review*, 18, 3-10.

Callon, M. (1986). Some elements of a sociology of translation: Domestication of the scallops and the fishermen of St Brieuc Bay. In J. Law (Ed.), *Power, Action And Belief: A New Sociology Of Knowledge?* (pp. 196-233). London: Routledge.

—. (1987). Society in the making: The study of technology as a tool for sociological analysis. In W. E. Bijker, T. P. Hughes, & T. J. Pinch (Eds.), *The Social Construction Of Technological Systems: New Directions In The Sociology And History Of Technology* (pp. 83-103). Cambridge, MA: MIT Press.

—. (1991). Techno-economic networks and irreversibility. In J. Law (Ed.), *A Sociology Of Monsters: Essays on power, technology and domination* (pp. 132-161). London: Routledge.
Castells, M. (1997). *The Power of Identity*. Malden, MA and Oxford: Blackwell Publishers.
—. (2004). An Introduction to the Information Age. In F. Webster, R. Blom, E. Karvonen, H. Melin, K. Nordenstreng, & E. Puoskari (Eds.), *The Information Society Reader* (pp. 138–149). London and New York: Routledge.
—. (2007). Communication, power and counter-power in the network society. *International Journal of Communication*, *1*, 238–266.
—. (2009). *Communication Power*. Oxford: Oxford University Press.
Castoriadis, C. (1975). *L'institution Imaginaire de la Société*. Paris: Éditions du Seuil.
Christiansen F. (2003). *Chinatown, Europe: An Exploration Of Overseas Chinese Identity In The 1990s*. London: Routledge.
Coen, E. (2004). *Contemporary tourism: diversity and change*. Amsterdam: Elsevier.
Coutard, O., & Guy, S. (2007). STS and the City: Politics and Practices of Hope. *Science, Technology, & Human Values*, *32*(6), 713-734.
Çınar, A., & Bender, T. (Eds.). (2007). *Urban Imaginaries: Locating the Modern City*. Minneapolis/London: University of Minnesota Press.
Çınar, A., & Bender, T. (2007a). The City: Experience, Imagination, and Place. In A. Çınar & T. Bender (Eds.), *Urban Imaginaries: locating the modern city* (pp. xi-xxvi). Minneapolis/London: University of Minnesota Press.
De Certeau, M. (1988). *The practice of everyday life*. Berkeley/Los Angeles: University of California Press.
De Sola Pool, I. (1983). *Technologies of Freedom*. Cambridge: Harvard University Press.
Falkheimer, J., & Jansson, A. (Eds.). (2006). *Geographies of communication: The spatial turn in media studies*. Göteborg: Nordicom.
Farías, I. (2011). Ensamblajes urbanos: la TAR y el examen de la ciudad. *Athenea Digital*, *11*(1), 15-40.
Farías I., & Bender, T. (Eds.). (2009). *Urban Assemblages. How Actor-Network Theory Changes Urban Studies*. London: Routledge.
Fathy, T. A. (1991). *Telecity: information technology and its impact on city form*. New York: Praeger.
Gelpi, A., & Tarantino, M. (2009). Chinatown Today: Schizophrenies of Italian Multicultural Publishing. In C. Giaccardi (Ed.), *Space, Media*

and Cultural Flows. Insights on Intercultural Communication (pp. 69-82). Milan: Vita e Pensiero.

Gibson, J. J. (1977). The theory of affordances. In R. E. Shaw & J. Bransford (Eds.), Perceiving, Acting, and Knowing. Hillsdale, NJ: Lawrence Erlbaum Associates.

—. (1979). The Ecological Approach to Visual Perception. Boston: Houghton Mifflin.

Gieryn, T. F. (2000). A Space for Place in Sociology. *Annual Review of Sociology*, *26*(1), 463-496

—. (2002). What Buildings Do. *Theory and Society*, *31*(1), 35-74.

Gitlin, T. (2003). *The Whole World is Watching: Mass Media in the Making and Unmaking of the New Left*. Berkley, Los Angeles and London: University of California Press.

Graham, S. (Ed.). (2004). *The Cybercities Reader*. London and New York: Routledge.

Graham, S., & Marvin, S. (1996). *Telecommunications and the City. Electronic spaces, urban places*. London and New York: Routledge.

Graham, S., & Marvin, S. (2001). *Splintering Urbanism. Networked Infrastructures, Technological Mobilities and the Urban Condition*. London and New York: Routledge.

Harvey, D. (1989). *The Condition of Postmodernity: An Enquiry into the Origins of Cultural Change*. Oxford: Basil Blackwell.

Hine, C. (2000). *Virtual Ethnography*. London: Sage.

Hommels, A. (2005). Studying Obduracy in the City: Toward a Productive Fusion between Technology Studies and Urban Studies. *Science, Technology, & Human Values*, *30*(3), 323-351.

—. (2005b). *Unbuilding Cities. Obduracy in Urban Sociotechnical Change*. Cambridge, MA: MIT Press.

Hommels, A. (2009). Changing Obdurate Urban Objects: the attempts to reconstruct the highway through Maastricht. In I. Farias & T. Bender (Eds.), *Urban Assemblages. How Actor-Network Theory Changes Urban Studies* (pp. 139-159). London: Routledge.

Humphreys, L. (2005). Cellphones In Public: Social Interactions In A Wireless Era. *New Media & Society*, *7*(6), 810-833.

—. (2010). Mobile Social Networks And Urban Public Space. *New Media & Society*, *12*(5), 763-778.

Jacobs, J. (1961). *Death and Life of Great American Cities*. New York: Vintage Books.

Jansson, A. (2005). *For a Geography of Communication*. Proceedings of the conference *Kulturstudier i Sverige. Nationell forskarkonferens* June 13-15 LINKÖPING, Sweden. Available online at

http://www.ep.liu.se/ecp_home/index.en.aspx?issue=015. Consulted on February 3, 2012.
Jenkins, H. (2006). *Convergence Culture*. New York: New York University Press.
Johnson-McGrath, J. (1997). Who Built the Built Environment? Artifacts, Politics, and Urban Technology. *Technology and Culture*, *38*(3), 690-696.
Kumar, K., & Makarova, E. (2008). The Portable Home: The Domestication of Public Space. *Sociological Theory*, *26*(4), 324-343.
Kirk, W. (1952). Historical Geography And The Concept Of The Behavioural Environment. *Indian Geographical Journal*, Silver Jubilee Edition, 152-160.
Latour, B. (1987). *Science in Action*. Milton Keynes: Open University Press.
—. (1988). *The Pasteurization Of France*. Cambridge, MA: Harvard University Press.
—. (1992). Where Are the Missing Masses? The Sociology of a Few Mundane Artifacts. In W. E. Bijker & J. Law (Eds.), *Shaping Technology/Building Society: Studies in Sociotechnical Change* (pp. 225-258). Cambridge, MA: MIT Press.
—. (1993). *We Have Never Been Modern*. Cambridge, MA: Harvard University Press.
—. (1999). On recalling ANT. In J. Law & J. Hassard (Eds.), *Actor network theory and after* (pp. 15-25). Oxford: Blackwell.
—. (2005). *Reassembling the Social: An Introduction to Actor-Network Theory*. Oxford, New York: Oxford University Press.
Law, J. (1987). Technology and heterogeneous engineering: The case of Portuguese expansion. In W. E. Bijker, T. P. Hughes, & T. J. Pinch (Eds.), *The social construction of technological systems: New directions in the sociology and history of technology* (pp. 111-134). Cambridge, MA: MIT Press.
Law, J. (1992). Notes on the Theory of the Actor Network: Ordering, Strategy and Heterogeneity. *Systems Practice*, *5*(4), 379-393.
Lefebvre, H. (1991). *The Production of Space*. Oxford: Basil Blackwell.
—. (1996). *Writings on Cities*. Translated and edited by E. Kofman and E. Lebas. Oxford: Blackwell.
Mahyar, A., & Meyersb, W. R. (2003). What is public about public space: The case of Visakhapatnam, India. *Cities*, *20*(5), 331–339.
Manzo, L. K. (2009). Spazialità emergenti, conflitti urbani contemporanei. Una ricerca etnografica nella Chinatown di Milano. In *Urban Hybridization In Contemporary Territories*. Available online at

http://www.urbanhybridization.net/indexfull.htm. Accessed on January 13th, 2012.
McCann, E. J. (1999). Race, Protest, and Public Space: Contextualizing Lefebvre in the U.S. City. *Antipode*, *31*(2), 163-184.
Mitchell, D. (1995). The End of Public Space? People's Park, Definitions of the Public, and Democracy. *Annals of the Association of American Geographers*, *85*(1), 108-133.
—. (1996). Introduction: Public space and the city. *Urban Geography*, *17*(29), 127-131.
Norman, D. A. (1988). *The Psychology of Everyday Things*. New York: Basic Books.
—. (1990). *The Design of Everyday Things*. New York: Doubleday.
Pascal, A. (1987). The vanishing city. *Urban Studies*, *24*, 597-603.
Pickvance, C. G. (1985). The Rise And Fall Of Urban Movements And The Role Of Comparative Analysis. *Environment And Planning D: Society And Space*, *3*, 31-53
Pinch, T. J., & Bijker, W. E. (1984). The Social Construction of Facts and artifacts: Or How the Sociology of Science and the Sociology of Technology Might Benefit Each Other. *Social Studies of Science*, *14*(3), 399-441.
Pinch, T. J., & Kline, R. (1996). Users as Agents of Technological Change: The Social Construction of the Automobile in the Rural United States. *Technology and Culture*, *37*(4), 763-795.
Pinch, T. J. (1996). Social Construction of Technology: A review. In R. Fox (Ed.), *Technological Change: Methods and Themes in the History of Technology* (pp. 17-36). Amsterdam: Harwood.
—. (1998). Theoretical Approaches to Science, Technology and Social Change: Recent developments. *Southeast Asian Journal of Social Science*, *26*, 7-16.
—. (2003). How Users and Non-Users Matter. In N. Oudshoorn & T. Pinch (Eds.), *How Users Matter: The Co-Construction of Users and Technology* (pp. 1-25). Cambridge, MA: MIT Press.
—. (2009). The Social Construction of Technology: The Old, the New, and the Nonhuman. In P. Vannini (Ed.), *Material Culture and Technology in Everyday Life: Ethnographic Approaches* (pp. 45-58). New York: Peter Lang.
—. (2010). On making infrastructure visible: putting the non-humans to rights. *Cambridge Journal of Economics*, 34(1), 77-89.
Russell, A. (2005). Myth and the Zapatista movement: exploring a network identity. *New Media & Society*, *7*(4), 559-577.

Salter, C. (1984). Urban Imagery and the Chinese of Los Angeles. *Urban Review*, 1, 15-20.

Sassen, S. (2001). *The global city: New York, London, Tokyo* (2nd Edition). Princeton: Princeton University Press.

Schuler, R. E. (1992). Transportation And Telecommunications Networks: Planning Urban Infrastructure for the 21st Century. *Urban Studies*, *29*(2), 297-309.

Shield, R. (1998). *Lefebvre, Love and Struggle: Spatial Dialectics*. London: Routledge.

Singer, B. D. (1970). Mass Media and Communication Processes in the Detroit Riot of 1967. *The Public Opinion Quarterly*, *34*(2), 236-245.

Soja, E. (1989). *Postmodern Geographies: The Reassertion of Space in Critical Social Theory*. London: Verso.

—. (1996a). The trialectics of spatiality, *Osterreichische Zeitschrift fur Soziologie*, *21*(2), 139-164.

—. (1996b). *Thirdspaces - Journeys into L.A. and Other Real and Imagined Places*. Oxford: Blackwell.

Sorkin, M. (1992). *Variations on a Theme Park: The New American City and the End of Public Space*. New York: Hill and Wang.

Stanek, L. (2005). The production of urban space by mass media storytelling practices: Nowa Huta as a case study. Paper presented at the conference *The Work of Stories*, Massachusetts Institute of Technology, 6-8 May.

Tarantino, M., & Tosoni, S. (2009a). Images of a Riot: Visual Representations And Symbolic Values in The Battle of Milan. In C. Giaccardi (Ed.), *Space Media and Cultural Flows: Insights on Intercultural Communication.* Milan: Vita e Pensiero.

Tarantino, M., & Tosoni, S. (2009b). The Battle of Milan: Social Representations of The April 12th Riots by two Italian Chinese communities. In G. Johanson, R. Smyth & R. French (Eds.), *Living Outside the Walls: the Chinese in Prato*. London: Cambridge Scholar Publishing.

Thrift, N. (2004). Driving in the City. *Theory, Culture & Society*, 21(4-5), 41-59.

Tosoni, S. & Tarantino, M. (2013), Media Territories And Urban Conflict: Exploring symbolic tactics and audience activities in the conflict over Paolo Sarpi, Milan. *International Communication Gazette*, 75(5-6).

Vale, L. J. (1995). The Imaging of the City: Public Housing and Communication, *Communication Research*, 22(6), 646-663.

Venturini, T. (2008). Piccola Introduzione Alla Cartografia Delle Controversie. *Etnografia E Ricerca Qualitativa*, 3. Available Online at htttp://www.medialab.sciences-po.fr/publications/Venturini-Introduzione_Cartografia_Controversie.pdf. Accessed February 3, 2012.

Webber, M. M. (1968). The Post City Age, *Daedalus*, *97*(4), 1091-1110.

CHAPTER II

THE CITY AS A MEDIUM OF MEDIA: PUBLIC LIFE AND AGENCY AT THE INTERSECTIONS OF THE DIGITALLY SHAPED URBAN SPACE

SEIJA RIDELL

1. Cracking urban infrastructures

In studying how digital and mobile-network technologies shape contemporary urban environments, the symbolic aspects of cities remain, without doubt, a prominent question[1]. People live their lives in fibre-optically networked, wirelessly connected and software-supported "cybercities" located in the midst of media representations, advertisements, signposts and other cultural artefacts. The presence of representations, in fact, has been augmented by digital technologies and contributes centrally to the constitution of the city's complex spatial-temporal texture.

However, the prevalent tendency–in cultural media studies, in particular–to focus on representations effectively obscures the ways in which the urban infrastructure conditions the phenomenal and symbolic dimensions of cities. Their increasingly digitalised and, thus, newly invisible material base and its entanglement with the networks of global capitalism have been discussed at length by critical geographers (see, for instance, Graham & Marvin, 1996, 2001; Graham, 2004b, c; Thrift, 2005; Crang & Graham, 2007; Kitchin & Dodge, 2011). Likewise, these scholars

---

[1] The chapter is a reworked version of Ridell's article "The Cybercity as a Medium", published in the *International Review of Information Ethics*, 12(3), pp. 12-20.

have pointed out how the production of space as part of urban activities becomes automatic as new devices are used and assimilated into everyday routines. In this way, people participate in the sinking of software beyond conscious recognition, where it becomes part of the self-evident background of daily existence (Thrift & French, 2002; van Kranenburg, 2008). One obvious reason for the effortless adoption of this taken-for-granted attitude towards technology in the urban context is that ordinary city dwellers lack access not only to the code-based "mechanosphere" of the city (Thrift & French, 2002) but also to even a rudimentary knowledge of its structure and functioning.

In my view, the studies of how computer code sorts and controls people's activities and interactions regardless of symbolic meanings are of crucial importance, particularly for the analyses of mass-media reception and of mobile uses of media in cities. More generally, the insights of critical spatial theorists provide fresh angles for conceiving the representational dimension of the "computable city" (Batty, 1997) in terms of power. Hence, I suggest in the chapter that theories of urban non-representation need to be brought into dialogue with symbolically biased views of culturally oriented media studies. The latter, for their part, are essential with regard to the critical analysis of contemporary city life for at least two major reasons.

To start with, it is not possible to tackle the city's non-representational "technological unconscious" (Thrift & French, 2002) without first representationalising it. Rendering perceptible the "values and ethical principles [inscribed] in the depths of the built information environment" (Star, 1999, p. 379) requires not only that we are aware of its presence but that we recognise the ways it supports, guides and constrains our behaviour. This cannot be done without configuring the non-representational in the realm of discourse–in other words, without translating its mute performative logic into systems of words and images. Hence, in order to get hold of the backstage, invisible technical machinery that enables and makes things happen on the urban front stage, we need to give it a symbolic form (cf. Star, 1999). This is because, as Mark Andrejevic (2009, p. 40) paraphrases Slavoj Žižek, the symbolic allows a distance between the code and what it defines, thus acknowledging "the possibility that things might be otherwise than how they 'directly' seem". The importance of this symbolic distance is in no way diminished by the fact that along with digitalisation, infrastructure not only has spread from the background all around us, but has also become attached to our bodies, almost literally becoming a part of it, moving with us when we travel in our cars, use our portable communication and media gadgets and carry around mundane consumer products tagged with a microchip.

Secondly, the representational dimension is indispensable if we want to ask how people lead their lives *together* in software-supported cities. This question frames the city as a public space and considers people in the urban context as public beings. To be *public*–both in the sense of visibility and of collectivity (Arendt, 1958; Weintraub, 1997)–our activities and interactions must take place in the symbolic dimension; they need to be performed visibly and/or audibly *in situ* or rendered otherwise in a way that is apparent to others.

In what follows, my point is not to prioritise the representational over the non-representational. Instead, I stress the pertinence of both in our attempts to understand the power dynamics of urban public life in the digitalised condition. We cannot grasp the "machine space" of present-day cities (Thrift & French, 2002) by merely analysing their symbolic dimension, nor is it possible to explain the specifics of contemporary urbanism by focusing exclusively on the city's invisible infrastructure. More precisely, then, I will direct my attention to points where the invisible materiality of the digitally sustained infrastructure meets, with intensity, the mediated and bodily lived phenomenality of the city. I do so because it seems to be at these intersections where the most intriguing questions for critical analysis emerge. This focus, for its part, requires that we take into account a third dimension, namely the *presentational*–in other words, the objects and technology-related performative acts that mediate between the *non-representational* and the *representational* in the urban context. Overall, the fundamental questions that arise at these intersections in the "triply articulated" cities concern people's active, though not necessarily self-reflexive, involvement in the dynamics of power, an involvement that has been characterised as post-hegemonic (see Lash, 2007) or posthuman (Hayles, 2006). In other words, what I wish to explore in the chapter is the technology-mediated conditioning of people's agency in contemporary urban environments.

2. Re-articulations of media and urban space

Essentially structured by the digitalisation and computerisation of its infrastructure, on the one hand, and by the pervasive presence of media technologies and representations, on the other, the present-day urban environment is composed of overlapping and interweaving spatialities of at least three analytically different kinds. We are concerned with a multilayered, spatial-temporal texture, a complex or a *medium* that is constituted by the co-existence and constant interpenetration of physical, discursive and virtual dimensions. The nesting of several spatialities makes the present-day city one of the richest (if not the richest) media environments or media ecologies, exemplifying what Matthew Zook and Mark Graham (2007, p. 468), following Henri Lefebvre, call the "hypercomplexity of space".

In this connection, a useful point of departure for a more general understanding of the term "medium" is to look at its philosophical definition as an ether, a gap, an interval or an in-between that enables human perception, communication, experience and understanding to take place, with each medium "giving us the world" in specific ways (Aristotle, 2000; Ridell & Väliaho, 2006). This inherently spatial notion of medium is given one prominent formulation in Marshall McLuhan's (1964) technologically inspired medium theory. According to McLuhan, media are environments that, rather than being mere technical means for conveying messages, actively wrap around us and also corporeally affect the ways we experience and (re)present ourselves and (inter)act with others. Combining this McLuhanesque notion of medium with the idea of the city as a huge, complex communication system "compatible with computers–and therefore media" (Kittler, 2006, p. 721) or as being in itself "the first supercomputer" (Strate, 2008, p. 138), it can be argued that the present-day city is, in a sense, the ultimate medium. Given the thorough digitalisation of infrastructure, the urban environment can be seen to incorporate every other medium, thereby accommodating the multiple spaces and overlapping spatial scales that constitute the city itself as a *medium of media*.

The notion of spatiality that I, in turn, wish to combine with this theoretical view of the medium derives its inspiration, in part, from Lefebvre's (1974/1991) and Edward Soja's (1985, 1996) conception of space as inherently social and relational: in other words, as something that is constituted and becomes constantly reproduced in the processes of human activity and interaction. There are also closer affinities, notably with Lefebvre's trialectical way of conceiving space through physical, mental and lived aspects. However, compared to his sweeping historical view that seeks to include modes of producing, perceiving, imagining and experiencing space, my focus in this chapter is confined to urban space as transformed or rearticulated through the development of communication and media technologies and, in particular, through their digitalisation–an issue that has only a minor role in both Lefebvre's and Soja's trialectics.[2] These technologies, as William J. Mitchell (2005, pp. 16-17) points out, while supplementing indirectly Lefebvre's (1991, p. 39) definition of representational space, also "add a dynamic layer of electronic information to the mise-en-scène established by an architectural setting and the meaningful objects and inscriptions that it contains" (see also Mitchell, 2003; Cuff, 2003; Gordon & Souza e Silva, 2011). Another difference between Lefebvre's spatial triad and the notion of multilayered space that I employ here is that the latter also conceives of (media) representations in terms of space, incorporating their specific discursive spatiality and its affordances.

While dwelling in and moving about the city, people continuously shift back and forth between the layers of the urban spatial constellation and simultaneously navigate between different spatial scales. Consider a person who sits on a bench in a city park connected wirelessly to the internet through her/his portable media device. S/he senses, at least subconsciously, the surrounding physical environment, noticing in an equally absent-minded manner the front covers of evening papers (metaphorically the porches of discursive spatiality) on nearby kiosk billboards; at the same time, within minutes, s/he also visits the webpage of an online newspaper, updates her/his status on Facebook and Twitter and then moves on to chat about politics on a web forum. Indeed, people's presence in the city today is increasingly a hybrid experience of being simultaneously present in many spatial and social environments (cf. Mitchell, 2003, pp. 28-29; Mitchell, 2005, p. 16; see also de Waal, 2008). Interaction in multiply interconnected spaces and, in particular, at their urban intersections also creates new kinds of social situations between people and challenges established codes of conduct. Thus, in the physical urban space, navigation between spaces often involves negotiation with social norms and rules, given that mobile technologies have rearranged the traditional boundaries between public and private territory, making it unclear how people should behave.[3]

---

[2] However, see Kirsch (1995) for a detailed discussion on applying Lefebvre's thinking in a technological context.

[3] One may notice also the increased possibilities of surveillance and control inherent in the digitally enabled interpenetration of diverse spatial layers. As Gordon and de Souza e Silva (2011, p. 2) point out, given the location awareness of mobile gadgets, we are constantly leaving behind data traces that create "a near

One interesting question at the intersection between the non-representational and the representational in the "media metropolis" (de Waal, 2008) concerns more specifically the simultaneous nesting and adaptability of spatial scales. First of all, digital-network technology and portable devices have made it possible for individuals to connect their personal "space bubbles" with various bigger, more or less remote groups of people and to do so while on the move. At the same time, portable gadgets can serve as delivery channels for mass-media representations, such as news, thus opening momentary, and often moving, micro spaces of globally spanned discursive publicness within the cityscape.

Second, the presence of mass media–both as materially palpable technologies (screens, printed materials, loudspeakers, billboards, panels) and as representations (the symbolic messages offered discursively on different media platforms)–in the city plays a major role in terms of scale. We can include here even the buildings of media houses, which not only furnish the local urban space but are organically connected to an invisible network made up of a global technostructure, like fungi and their rhizomes. The animated façades covering the entire surfaces of buildings–their outer skins–likewise grow out of the digitalised media machinery that produces and displays the moving images (cf. Jacob, 2011, p. 153; Krajina, 2009, p. 406; also McQuire, 2008, pp. 126–127). Regarding distinct mass-media technologies, such as television, their specific feature is the ability to adapt scales between physically distant locales both technologically and discursively. A site-specific television set in a local pub, for example, can transmit live audiovisual narrative from the other side of the globe, thereby bringing together, in one communicative presence, the local "here" and the distant "there" (McCarthy, 2001).[4] The levels of non-representation, presentation and representation are involved because neither the digitally embedded television technology nor the televisual representation alone are able to accomplish this bridging of scales.

One way to illustrate the scale-adapting presence of mass media in cities is by extending the notion of "double articulation", introduced in the early 1990s for the ethnographic study of domestic media consumption. As pointed out by Roger Silverstone (1994), not only do media carry

---

comprehensive map of where we are in relation to everything else" (see also Sheller & Urry, 2006, p. 222).

[4] Cf. Morley (1994, 1996) on television's role in connecting the household with the wider social life that is physically located outside its walls.

symbolic messages, they are themselves–as material devices and objects–imbued with cultural meaning (see also Silverstone & Hirsch, 1992). A unique feature of media as objects is precisely their liminality: they both mediate messages and act as intermediaries between different and differently scaled worlds. This intermediary position is a site of struggle because it is through access to people's homes and urban spaces that media corporations and other powerful agencies attempt to insinuate images and discourses into both the household's and the city's daily rhythms and routines.

In contemporary urban settings, the meaningful materiality (and the material meaningfulness) of both mass and personalised media forms a junction where the digitalised technostructure and the phenomenality of the city are able to interface with each other in multiple specific ways. Exploration of these intersecting (re-)articulations may uncover, after closer scrutiny, the conditioning efficacy of the software-supported urban infrastructure and also help us grasp more lucidly the ways in which the pervasive presence of media–both as doubly articulated objects and as representations–constitutes the contemporary city itself as a medium.

It is pertinent to notice that from a medium-theory starting point, what we essentially should focus on when studying the role of media in the urban context is, in fact, presentation, that is those material objects and displaying gestures that make things perceptible to other city dwellers, enabling their public sharing. In other words, it is through the intermediary capacity of the presentational that the non-representational is articulated in the first place and through which the representational gains its perceivable existence. This, then, makes apparent the need to talk about the media–given the digitalised urban condition–as (at least) *triply articulated* instead of seeing their articulation as merely double, an issue that deserves a separate theoretical discussion.[5]

---

[5] Despite his hesitation to talk about more than two articulations of media, Silverstone (2006, p. 239) nevertheless points indirectly to three dimensions in the way that ICTs articulate in stating that as objects, these technologies are both material and symbolic; the third dimension consists of the "contents" that the technologies carry. For an explanation regarding the notion of triple articulation that focuses on media use in the domestic context, see Hartmann (2006) and Courtois et al. (2012).

3. Conditions of public life in the digitalised city

Obviously, urban public life has transformed quite drastically due to the increasing digitalisation of material infrastructures. At the same time, sharing in physical space has grown increasingly complex in contemporary, spatially multilayered and multi-scaled cities. One pressing issue that this development has brought to the fore concerns the power to condition people's activities and interactions. When tackling the conditions of urban agency, discussions on "post-hegemonic power" (Lash, 2007) and "posthuman subjectivity" (Hayles, 2006; Gane, Venn, & Hand, 2007) appear especially relevant and fruitful because they allow for studying power dynamics beyond the level of urban representation.

To begin with, in medium-theory terms, a central issue of power in contemporary cities concerns the rhetoric of non-representationally supported and presentationally staged public urban space. The city, as a triply articulated medium, interpellates the dwellers and *flãneurs* to specific subject positions at all layers of the spatial constellation, making it particularly pertinent to ask what forms and strategies the hailing assumes at points where the infrastructure and the phenomenal-symbolic dimension of the city intersect. At these densely mediated intersections, one urgent question is how software regulates and controls the kinds of public agency that are rhetorically on offer in urban milieus. Another important issue is concerned with the ethical and political implications of the overwhelming presence of mass media. It can be suggested that the rhetoric of the city medium is at its most persuasive at sites where mass communication technologies–as digitally embedded presentational objects–furnish urban spaces and simultaneously put forward symbolic messages specifically for public display. Anna McCarthy remarks, for example, that the positioning of television screens in urban spaces is carefully planned in order to standardise certain patterns of movement and perception for users who pass through those locations. At the same time, commercial practices that guide the production and circulation of texts and programme forms work to "commodify the spectator's position in space for sale to advertisers" (McCarthy, 2001, pp. 11-12).

The pervasiveness of advertising in particular constitutes the present-day urban environment itself much like a mass medium with its one-way patterns of communication. This can be observed most clearly in shopping malls, which not only have television screens scattered all over but which also mimic television, in that "the shopper strolls through experiences as he or she might scan through TV channels" (Goss, 1993, p. 39). Leif Dahlberg (2006, p. 41) argues critically that "mediated discourse in (physical) public places in contemporary western society to a large extent has been monopolised by an advertising monologue that *shuts out* other public discourses" and speaks to us "as consumers". Anne Cronin (2006a, p. 627) also importantly points out that the persuasive textuality of the commercialised city is strongly embodied, in that "advertising attempts to target and inhabit everyday commuting and shopping routes and become part of the fabric of people's urban experience" (see also Krajina, 2009, p. 410, 415).

It should be stressed, however, that space as such is not an external frame for human action, but people's activities and interactions do contribute to the production and maintenance of space and its specific characteristics. In terms of the dynamics of power in urban space, then, it is not enough to analyse how the mass-media dominated urban rhetoric addresses city dwellers, and as what kinds of subjects it addresses them; equally importantly, it should be considered how the interpellation is actually received. Do people accept the position of consumer audience that is offered them, or can we find acts and practices that constitute aberrant and resisting decodings (Hall, 1980)? Also, to paraphrase Jon Goss (1993), McCarthy (2001) and Cronin (2006a), who all apply Michel de Certeau's (1980/1984) well-known distinction, how are the strategies of the spatially powerful met by the tactics of ordinary city dwellers in actual practice?

Another pertinent power-related issue concerns people's technology and media-related activities in contemporary cities–specifically, the ways that personalised uses of media contribute to the city medium's mode of address and mould the city itself as a spatial constellation.[6] Studies on the use of portable communication and media devices in the city, such as headphones and music players (mp3 players, iPods), laptops and the increasingly multi-functional mobile phones, report that these technologies tend to separate people who share physical urban space. For example, in their study on wireless internet use in paid and free Wi-Fi cafes, Keith

[6] Here, a more representation-oriented set of research questions could be formulated by focusing on audience activities in the media-saturated city (see Ridell, 2010). In the present-day urban space, people not only receive the strategic hailing by the broadcasting and commercial display screens and panels, but are also constantly acting as an audience for other people's performances that relate to uses of handheld media devices, such as mobile phone conversations (see e.g. Drotner, 2005).

Hampton and Neeti Gupta (2008) observed similarities between the use of mobile phones and laptops, in that both are often employed as portable "involvement shields", as theorised by Erving Goffman (1963), to avoid contact with co-located strangers and for socialising instead with remote but already familiar people and groups (see also Ito, Okabe, & Anderson, 2009; Hampton, Livio, & Sessions, 2009). On the whole, mobile technologies are seen to construct individualised space capsules, bubbles or "telecocoons" with invisible barricades around them that co-exist in their separation on the urban stage dominated by the pervasively furnishing presence of mass media. These technologies can even be described as "territory machines" that temporarily, but repeatedly, seize and appropriate physical urban space for (inter)personal purposes. Whether the uses of portable communication and media devices are "defensive or offensive postures" (de Waal, 2008), they strengthen a tendency in the physical urban space that Michael Bull (2004, p. 278) calls "public privacy" and what Hampton and Gupta (2008, p. 835) refer to as "public privatism".

Studies on the uses of mobile media demonstrate that the contemporary urban environment is without doubt public in the first sense of Hannah Arendt's (1958) dual definition: people in their telecocoons appear visibly and audibly to each other, sometimes even exposing their most intimate thoughts and deeds and exhibiting their community networks to an embarrassing extent. However, being perceptible to others and witnessing others' performances in a space does not automatically make that space public in the collective sense, as recognised by Arendt. In other words, a collectively public space is not merely a platform of appearance but refers simultaneously to a space that provides opportunities to address and encounter previously unknown and not necessarily like-minded others regarding matters of public concern (Barnett, 2004, pp. 406-407). Marcel Hénaff and Tracy Strong (2001, p. 1) formulate the collective aspect in a slightly differently way, asserting that we can call public any space "in which human beings encounter each other with the intention of determining how their lives in common should be lived".

In light of recent research, it certainly seems that the use of handheld devices in urban space discourages interacting with physically co-present strangers and initiating discussions with them on issues that exceed either an individual's personal or the in-group's aspirations. There may be publicly oriented activities involved, but these tend to take place online instead of on the physical layer of urban environment, where users more often seem to step into the role of audience, on the one hand, and that of a community member, on the other. However, there are also occasions–as in the case of Hampton and Gupta's (2008) "placemakers", with their laptops–when people use personal media devices to initiate, rather than shut out, interaction with unfamiliar others around them. This calls for further theorisation, as well as empirical studies in urban locations, in order to explore whether and how "smart" mobiles, compared to portable media that are more traditional, such as newspapers or books, serve as facilitators not only of community-oriented interpersonal communication but also of interactive engagements between people as publics (see Hampton, Livio, & Sessions, 2009).[7]

4. Contesting post-hegemonic power?

At a first glance, the mass-media saturation of contemporary cities suggests that the dominant mode of urban spatial address cannot be easily challenged. We may wonder whether alternative decodings are even possible in thoroughly commercial spaces like shopping malls, for example, where all forms of interpretation, including the resistant ones, have been anticipated and assimilated as parts of the seamlessly seductive urban phenomenality (cf. Allen, 2006).

In actual practice, however, it is fairly easy to imagine ways of politicising and countering the consumerist rhetoric of the city medium at the level of representation. This is exactly what interventionist urban art projects, critical consumer campaigns and other "subvertising" activities attempt to do (see Dahlberg, 2006; Cronin, 2006a, b). The much trickier question concerns the habitual use of both mass and personalised media in urban settings. How does one render problematic the taken-for-granted nature of urban media routines and then challenge their power-relatedness; in other words, how does one make visible and politicise the ways that

[7] One case for such an empirical study is the recent Occupy movement, in which new media technologies are used, on the one hand, to organise and coordinate collective activities in the physical public space in different cities around the world and, on the other hand, to make these activities globally visible in online spaces. From the perspective suggested in this chapter, it would be interesting to study systematically whether the physical gatherings extend beyond more or less closed community participation and move towards fostering communication in which people assume, on the spot, the role of the public (for a recent case study, see Gerbaudo, 2012). For an action theory-based discussion on the differences between the positions of audience, community member and public, see Ridell (2012).

people, through their own activities, contribute to the fragmentation and compartmentalisation of city space at the same time that they participate in confirming the invisible infrastructural conditions of urban public life?

If we are to accept Scott Lash's (2007) view of post-hegemonic power, it is indeed not possible to challenge the conditioning of urban public life and agency that takes place at the infrastructural level of non-representation. This is because, following Lash (2007), the routines involved are firmly located on the ontological level, residing in the "algorithmic, *generative* rules" that "we do not encounter [...] in the same way that we encounter constitutive and regulative rules" (p. 71). These generative rules pervade our entire social and cultural life, but because they "have to do with the thing-itself that is never encountered" (ibid.), their critical reflection seems to be out of reach. Hence, the implication is that in the context of the sinister post-hegemonic power, resistance is not only impossible and futile but also irrelevant. In fact, as domination now constitutes us from the inside (p. 61), grasping "us in our very being" (p. 75) and having made the brain "immanent in the system itself" (p. 60), we have become our own (post)hegemons, which seems to leave destructive self-revolt as the only option for resistance.

In contrast to this paralysing scenario, I wish to suggest that there are ways to challenge and resist the post-hegemonic dynamics of urban spatial power at the level of routinised action. Paradoxically enough, digital technologies themselves offer abundant and partly unexplored opportunities for venturing into the elusive urban infrastructure and, subsequently, for reflecting on our "conspiratorial" involvement in its performance. Embryonic forms of such ventures can be found here and there–in mobile gaming that explores the surveillance of urban areas or that plays with the "seams" in the wireless-network design, through planned sousveillance activities and, one more illustration, by disclosing attacks by "street-level internet crackers", to name a few examples (see Sweeny & Patton, 2009; Chalmers et al., 2005; Mann, Nolan, & Wellman, 2003; Graham, 2004a).

Tactics for rendering discernible and for problematising the routinely unrecognised in the contemporary urban setting require treatment beyond the scope of this chapter. What is important to note, however, is the pivotal role of the symbolic or representational dimension as an integral part of subversive action. To start with, in order to take a reflexive stance towards the software-supported infrastructure and to ponder critically the nature of our compliance with it, we need to recognise where and how that structure exists.[8] Once we do so, the gestures themselves that politicise and disrupt how the infrastructural system works, both independently of and through people's technology and media -related activities, are plentiful and, at times, even trivial.[9]

In other words, configuring the digitally embedded infrastructure in the realm of representation is the necessary first step in the process of coming to grips with it in terms of power. A central part of sousveillance activities, for example, is to uncover the disappearance of digital technologies into the fabric of buildings, objects and bodies and to make the technology-mediated surveillance known to others by means of photographing, videotaping or evoking counter-performances (Mann, Nolan, & Wellman, 2003). Urban internet crackers, for their part, with their location-aware handhelds track "points from which they can enter the local, broadband wireless networks of corporations". Once connected to these hotspots of network coverage, the "street-level activists mark up their boundaries with chalk so that these hidden infrastructures can be publicly consumed" (Graham, 2004a, p. 16).

One more example of countering post-hegemonic power by means of representation is critical RFID[10] activists' attempts to expose–through public boycotts and protests both offline and online–the use, and future potential, of "spychips" as a corporate and governmental surveillance tool.[11] A crucial theoretical question to be posed in connection with this is whether it would make a difference regarding the conditioning of the digitally reconfigured posthuman agency if people were more aware of the ubiquitous "cognisphere" that invisibly surrounds them in the contemporary urban environment and, furthermore, if they realised how

---

[8] As David Beer (2009, p. 1000) points out in the context of web-based social media, one of the pressing questions about the nature of power today concerns the fact that "we simply do not understand how the material infrastructures of Web 2.0 play out in the lives of individual users, how the software constrains and enables, how it formulates hierarchies, shapes the things people encounter, and so on". See also Ratto (2007, pp. 24-25).

[9] The vulnerability of the technosystem, along with the huge economic interests that drive its development and maintenance, is, of course, one of the main reasons for keeping it hidden from the majority of people (cf. Star & Bowker, 2002, pp. 237-238).

[10] RFID is an acronym for Radio Frequency Identification and refers to a microchip technology that is used to track items at a distance. Tiny RFID chips are embedded in all kinds of consumer products, from razors to clothes to cars.

[11] See www.spychips.com/

they themselves contribute to it through their technology-related activities (cf. Hayles, 2006; Gane, Venn, & Hand, 2007; Ratto, 2007). As RFID activist Katherine Albrecht (2005) emphasises, in order for people to agree to bear RFID-tagged clothes and actively submit to being tracked, they must first be aware that these tags exist (see also Albrecht & McIntyre 2005). In other words, if people have no idea that their everyday objects and devices contain an RFID chip and what this implies in terms of their own activities, how can they think–much less do something–about the ways these tags are connected to databases and other "sorting software"?

Equally important, politicising and challenging the post-hegemonic dynamics power in the urban context also entails ventures into the embodied texture of contemporary cities where private space bubbles exist and move about in close proximity while still ignoring one another's presence in a more or less calculated manner. Disturbing the self-evident separation of telecocoons and rendering problematic their offensive variants would make palpable, and raise questions about, the colonisation of urban space by domesticating and customising desires. This, again, may open up opportunities for countering the forces that work to deprive urban public space of its collective significance. However, keeping in mind that courteously acknowledging other people's presence while simultaneously declining interaction with them is, in fact, a *sine qua non* of city life (Lofland, 1989, p. 462)[12] makes the actual form of such interventions–and research about them–an urgent topic for further critical thinking and discussion.

References

Albrecht, K. (2005). Dismantling the RFID Journal's Critique of *Spychips.* http://www.spychips.com/book/roberti-rebuttal.html.

Albrecht, K., & McIntyre, L. (2005). *Spychips*. New York: Plume.

Allen, J. (2006). Ambient Power: Berlin's Potsdamer Platz and the Seductive Logic of Public Spaces. *Urban Studies*, 43(2), 441–455.

Andrejevic, M. (2009). Critical Media Studies 2.0: An Interactive Upgrade. *Interactions: Studies in Communication and Culture*, 1(1), 35–51.

[12] In the urban context, this "civil inattention", in Goffman's (1963, pp. 83-88) terms, is one guarantee for sensibly managing the constant flow of fleeting encounters with large numbers of unknown persons and groups.

Arendt, H. (1958). *The Human Condition*. Chicago: University of Chicago Press.

Aristotle. (2000). *Aristotle in Twenty-Three Volumes. Issue 8: On the Soul, Parva Naturalia, On Breath*. Translated by W. S. Hett. Loeb Classical Library. Cambridge: Harvard University Press.

Barnett, C. (2004). Convening Publics: The Parasitical Spaces of Public Action. In K. Cox, M. Low, & J. Robinson (Eds.), *The Sage Handbook of Political Geography* (pp. 403–417). Los Angeles: Sage Publications.

Batty, M. (1997). The Computable City. *International Planning Studies*, 2, 155–173.

Beer, D. (2009). Power Through the Algorithm? Participatory Web Cultures and the Technological Unconscious. *New Media & Society*, 11(6), 985–1002.

Bull, M. (2004). "To Each Their Own Bubble": Mobile Spaces of Sound in the City. In N. Couldry & A. McCarthy (Eds.), *Mediaspace. Place, Scale and Culture in a Media Age* (pp. 275–93). London & New York: Routledge.

Chalmers, M., Marek, B., Brown, B., Hall, M., Sherwood, S., & Tennent, P. (2005). Gaming on the Edge: Using Seams in Ubicomp Games. In The Proceedings of the International Conference on Advances in Computer Entertainment Technology, Valencia, June 15–17, pp. 306–309. http://eprints.gla.ac.uk/3423/1/Gaming_on_the_edge.pdf.

Courtois, C., Verdegem, P., & De Marez, L. (2012). The Triple Articulation of Media Technologies in Audiovisual Media Consumption. *Television & New Media*, 20(6), 1–19.

Crang, M., & Graham, S. (2007). Sentient Cities: Ambient Intelligence and the Politics of Urban Space. *Information, Communication & Society*, 10(6), 789–817.

Cronin, A. M. (2006a). Advertising and the Metabolism of the City: Urban Space, Commodity Rhythms. *Environment and Planning D: Society and Space*, 24, 615–632.

—. (2006b). Urban Space and Entrepreneurial Property Relations: Resistance and the Vernacular of Outdoor Advertising and Graffiti. On-line paper published by the Department of Sociology, Lancaster University, UK. http://www.comp.lancs.ac.uk/sociology/papers /Cronin-advertisingandgraffiti.pdf

Cuff, D. (2003). Immanent Domain: Pervasive Computing and the Public Realm. *Journal of Architectural Education*, 43–49.

Dahlberg, L. (2006). On the Open and Closed Space of Public Discourse. *Nordicom Review*, 27(2), 35–52.

de Certeau, M. (1984). *The Practice of Everyday Life*. Berkeley: University of California Press. (French original *Arts de faire*, 1980)

de Waal, M. (2008). Locative media and the City. From BLVD-urbanism Towards MySpace urbanism. http://www.martijndewaal.nl/?p=116

Drotner, K. (2005). Media on the Move: Personalised Media and the Transformation of Publicness. In S. Livingstone (Ed.), *Audiences and Publics: When Cultural Engagement Matters for the Public Sphere* (pp. 187–211). Bristol: Intellect.

Gane, N., Venn, C., & Hand, M. (2007). Ubiquitous Surveillance: Interview with Katherine Hayles. *Theory, Culture & Society*, 24(7–8), 349–358.

Gerbaudo, P. (2012). *Tweets and the Streets: Social Media and Contemporary Activism*. London: Pluto Press.

Goffman, E. (1963). *Behavior in Public Places. Notes on the Social Organization of Gatherings*. New York: Free Press.

Gordon, E., & de Souza e Silva, A. (2011). *Net Locality: Why Location Matters in a Networked World*. West Sussex: Wiley-Blackwell.

Goss, J. (1993). The "Magic of the Mall": An Analysis of Form, Function, and Meaning in the Contemporary Retail Built Environment. *Annals of the Association of American Geographers*, 83(1), 18–47.

Graham, S. (2004a). Introduction: From Dreams of Transcendence to the Remediation of Urban Life. In S. Graham (Ed.), *The Cybercities Reader* (pp. 3–29). London: Routledge.

—. (2004b). Beyond the "Dazzling Light": From Dreams of Transcendence to the "Remediation" of Urban Life: A Research Manifesto. *New Media & Society*, 6(1), 16–25.

—. (2004c). The Software-sorted City: Rethinking the "Digital Divide". In S. Graham (Ed.), *The Cybercities Reader* (pp. 324–332). London: Routledge.

Graham, S., & Marvin, S. (1996). *Telecommunications and the City. Electronic Spaces, Urban Places*. London and New York: Routledge.

Graham, S., & Marvin, S. (2001). *Splintering Urbanism: Networked Infrastructures, Technological Mobilities, and the Urban Condition*. New York: Routledge.

Hall, S. (1980). Encoding/decoding. In S. Hall, D. Hobson, A. Lowe, & P. Willis (Eds.), *Culture, Media, Language. Working Papers in Cultural Studies 1972–1979* (pp. 128–138). London etc.: Hutchinson.

Hampton, K. N., & Gupta, N. (2008). Community and Social Interaction in the Wireless City: Wi-fi Use in Public and Semi-public Spaces. *New Media & Society*, 10(6), 831–850.

Hampton, Keith, Oren Livio, and Lauren Sessions. 2009. "The Social Life of Wireless Urban Spaces: Internet Use, Social Networks, and the Public Realm." Paper presented at the pre-conference workshop at the International Communication Association (ICA) Conference Chicago, Illinois, May 20–21. http://www.lirneasia.net/wp-content/uploads/2009/05/final-paper_hampton_et_al.pdf

Hartmann, M. (2006). The Triple Articulation of ICTs. Media as Technological Objects, Symbolic Environments and Individual Texts. In T. Berker, M. Hartmann, Y. Punie, & K. Ward (Eds.), *Domestication of Media and Technology* (pp. 80–102). Open University Press, Maidenhead.

Hayles, K. (2006). Unfinished Work: From Cyborg to Cognisphere. *Theory, Culture & Society*, 23(7–8), 159–166.

Hénaff, M., & Strong, T. B. (2001). Introduction: The Conditions of Public Space: Vision, Speech and Theatricality. In M. Hénaff, & T. B. Strong (Eds.), *Public Space and Democracy* (pp. 1–31). Minneapolis: University of Minnesota Press.

Ito, M., Okabe, D., & Anderson, K. (2009). Portable Objects in Three Global Cities. In R. Ling, & S. W. Campbell (Eds.), *The Reconstruction of Space and Time: Mobile Communication Practices* (pp. 67–87). New Brunswick: Transaction Publishers.

Jacob, A. (2011). Digital Media and Architecture. An Observation. In M. Rieser (Ed.), *The Mobile Audience, Media Art and Mobile Technologies* (pp. 141–154). Amsterdam: Rodopi.

Kirsch, S. (1995). The Incredible Shrinking World? Technology and the Production of Space. *Environment and Planning D: Society and Space*, 13, 529–555.

Kitchin, R., & Dodge, M. (2011). *Code/Space Software and Everyday Life*. Cambridge, Mass. & London: The MIT Press.

Kittler, F. A. (1996). The City is a Medium. *New Literary History*, 27(4), 717–729.

Krajina, Z. (2009). Exploring Urban Screens. *Culture Unbound*, 1, 401–430.

Lash, S. (2007). Power after Hegemony. Cultural Studies in Mutation? *Theory, Culture & Society*, 24(3), 55–78.

Lefebvre, H. (1991). *The Production of Space*. Oxford: Blackwell Publishing. French original *Production de l'espace*, Anthropos, 1974.

Lofland, L. H. (1989). Social Life in the Public Realm. *A Review Journal of Contemporary Ethnography*, 17(4), 453–482.

Mann, S., Nolan, J., & Wellman, B. (2003). Sousveillance: Inventing and Using Wearable Computing Devices for Data Collection in Surveillance Environments. *Surveillance & Society*, 1(3), 331–355.
McCarthy, A. (2001). *Ambient Television: Visual Culture and Public Space*. Durham & London: Duke University Press.
McLuhan, M. (1964). *Understanding Media: The Extensions of Man.* New York: McGraw Hill.
McQuire, S. (2008). *The Media City: Media, Architecture and Urban Space*. Los Angeles etc.: SAGE.
Mitchell, W. J. (2003). *Me++: The Cyborg Self and the Networked City*. Cambridge, Mass.: The MIT Press.
—. (2005). *Placing Words: Symbols, Space, and the City*. Cambridge, Mass.: The MIT Press.
Morley, D. (1994). Between the Public and the Private: The Domestic Uses of Information and Communications Technologies. In J. Cruz & J. Lewis (Eds.), *Viewing, Reading, Listening* (pp. 101–123). Boulder etc.: Westview Press.
—. (1996). The Geography of Television: Ethnography, Communications, and Community. In J. Hay, L. Grossberg, & E. Wartella (Eds.), *The Audience and Its Landscape* (pp. 317–342). Boulder, Col.: Westview Press.
Ratto, M. (2007). Ethics of Seamless Infrastructures: Resources and Future Directions. *IRIE—International Review of Information Ethics*, 8, 21–27.
Ridell, S. (2010). Audiencing (in) the media city. Paper presented at ECREA's 3rd European Communication Conference, Hamburg, Oct 12–15.
—. (2012). Mode of Action Perspective to Engagements with Social Media. Articulating Activities on the Public Platforms of Wikipedia and YouTube. In H. Bilandzic, G. Patriarche, & P. J. Traudt (Eds.), *The Social Use of Media: Cultural and Social Scientific Perspectives on Audience Research* (pp. 17–35). Bristol, UK & Chicago, USA: Intellect.
Ridell, S., & Väliaho, P. (2006). Mediatutkimus käsitteiden kudelmana (Media Studies as a Fabric of Concepts). In S. Ridell, P. Väliaho, & T. Sihvonen (Eds.), *Mediaa käsittämässä* (Grasping Media) (pp. 7–26). Tampere: Vastapaino.
Sheller, M., & Urry, J. (2006). The New Mobilities Paradigm. *Environment and Planning A*, 38, 207–226.
Silverstone, R. (2006). Domesticating Domestication. Reflections on the Life of a Concept. In T. Berker, M. Hartmann, Y. Punie, & K. Ward

(Eds.), *Domestication of Media and Technology* (pp. 229–248). Open University Press, Maidenhead.
—. (1994). *Television and Everyday Life*. London: Routledge.
Silverstone, R., & Hirsch, E. (Eds.). (1992). *Consuming Technologies: Media and Information in Domestic Spaces*. London: Routledge.
Soja, E. W. (1985). The Spatiality of Social Life: Towards a Transformative Retheorisation. In D. Gregory, & J. Urry (Eds.), *Social Relations and Spatial Structures* (pp. 90–127). New York: St. Martin's Press.
—. (1996). *Thirdspace. Journeys to Los Angeles and Other Real-and-Imagined Places*. Oxford: Blackwell.
Star, S. L. (1999). The Ethnography of Infrastructure. *American Behavioral Scientist*, 43(3), 377–391.
Star, S. L., & Bowker, G. (2002). How to Infrastructure. In L. Lievrouw & S. Livingstone (Eds.), *The Handbook of New Media* (pp. 151–162). London: Sage.
Strate, L. (2008). Studying Media *as* Media: McLuhan and the Media Ecology Approach. *Media Tropes eJournal*, 1, 127–142.
Sweeny, R., & Patton, P. (2011). CitySneak: Play, Pedagogy, and Surveillance. In A. de Souza e Silva & D. Sutko (Eds.), *Digital Cityscapes. Merging Digital and Urban Playspaces* (pp. 204–216). New York: Peter Lang.
Thrift, N. (2005). *Knowing Capitalism*. London: Sage Publications.
Thrift, N., & French, S. (2002). The Automatic Production of Space. *Transactions of the Institute of British Geographers*, 27(3), 309–335.
van Kranenburg, R. (2008). *The Internet of Things. A Critique of Ambient Technology and the All-seeing Network of RFID*. University of Amsterdam: The Institute of Network Cultures and Notebook 02.
Weintraub, J. (1997). The Theory and Politics of the Public/Private Distinction. In J. Weintraub, & K. Kumar (Eds.), *Public and Private in Thought and Practice. Perspectives on a Grand Dichotomy* (pp. 1−42). Chicago & London: The University of Chicago Press.
Zook, M., & Graham, M. (2007). Mapping DigiPlace: Geocoded Internet Data and the Representation of Place. *Environment and Planning B: Planning and Design*, 34, 466–482.

CHAPTER III

THE BATTLEFIELD OF URBAN BRANDING IN FOOTBALL MEDIA SPECTACLES

SAMI KOLAMO

Introduction

> The true legacy of this spectacle will be in our ability to showcase South African and African hospitality and humanity–to change once and for all perceptions of our country and our continent among peoples of the world. That depends on all of us; and to that we can attach no price! (The deputy-president of South Africa, Kgalema Motlanthe, cited in Ngonyama, 2010, p. 176)

During the South African World Cup (2010) FIFA (*Fédération Internationale de Football Association*), the owner of the rights to this global sports product, together with its close allies (the Official Partners, Local Organising Committee, South African Football Association and Host City authorities) adopted a "total approach" to the branding of the local urban space, involving not only a whole repertoire of slogans, logos, liveries, official merchandise, public relations, urban revitalization projects, flamboyant design statements, but also comprehensive support from locals (Mayes, 2008; Ward, 1998, p. 1). In essence, all was planned to look attractive before the television cameras, in order to reverse the "Afro-pessimism" of the global media through attestations that "Africans can do it", and thereby mediate a positive image of South Africa's hospitable, "all-inclusive world-class cities" to the rest of the world. The image-building process of (South) Africa's new hospitality society was in line with FIFA's interest in maximising profit by offering ideal audiences for advertisers. Importantly, the marketers of Renaissance (South) Africa did not seem to understand places as spatial entities that are always rife with internal differences, conflicts and multiple definitions of identity (e.g.

Massey, 2005), but attempted to simplify them as resources for capital accumulation.

In this article, I take under scrutiny the phenomenon of urban branding in terms of place promotion, and the significance of local people's performative acts within it. Starting with specific branded enclaves in urban space, the power of the (mass) media to organise social space is emphasized. In the subsequent two sections, the focus is on the local people. At first, I shed light on how a "carnival-like" attitude among fans was purposefully promoted as being an ideal feeling that everyone should express. Next, I consider the transgressive acts of fans and the political resistance of activists in this specific case. Turning focus on the practises of not-so-well tamed fans and activists reveals urban branding in a context-dependent and inherently controversial spatial practice. At the same time, the logic of the (mass) mediated spectacle is illustrated, in particular, the ways it shifts the attention of the people away from unequal material conditions and social problems (Debord, 1983; Crary, 2000).

In the concluding section, by introducing the concept of the all-inclusive world-class city, the strategies of urban branding in the context of the football media spectacle are criticised against an ethically and politically sensitive framework. The article's arguments are based on the author's fieldwork in South Africa and material from Finnish television broadcasts, as well as on critical works and comments by researchers, journalists and activists on the South African World Cup[1].

---

[1] During the one-week fieldwork period the author collected research material by observation, writing diary notes and taking photographs. In South Africa the target cities were Johannesburg, the biggest metropolis in the country, and Rustenburg, which has about 400.000 inhabitants. In these cities, in addition to stadiums and their surroundings, the author studied the Public Viewing arenas set up by FIFA, plus other urban spaces essential in a consideration of the movements of tourists, fans and the visible public (city centres, shopping centres, market places and other main squares, airports, train and bus stations and road-sides). In order to make a comparison, the author also visited the district of Soweto, which is populated by black people, and in particular went to the 'unofficial' tournament studios set up in the homes of the local inhabitants. The research material has been supplemented with other essential material, especially in the sections dealing with the resistance of activists. This is because both in the mainstream media and in the "cleansed" and controlled urban spaces, critical statements regarding the tournament arrangements and the alternative voices of the fan culture had effectively been silenced.

1. Branding urban space through (mass) mediated enclaves

One of the most controversial construction projects of the South Africa World Cup was the new ultramodern Green Point Stadium (later named Cape Town Stadium) in front of the famous Table Mountain in Cape Town. The city officials originally planned to use the already existing Athlone Stadium, located in a working-class neighbourhood, to improve the weak employment situation in the district and develop its services. However, according to representatives of FIFA, "housing in Athlone and the overall landscape of the area would not form a suitable background for television viewers" (Hlatswayo & Blake, 2011, pp. 227-229). For the television viewers, FIFA wanted to offer spectacular scenery, not shacks and images of large-scale poverty. To make way for the new stadium and "televisualized scenery" the local poor were settled into an inhumane tin can town (Kolamo & Vuolteenaho, 2011, p. 23). The case is an example par excellence both of FIFA's ability to impose its own interests on elected governments and, above all, the "sacred" status of media publicity and the related advertising revenues from staging football spectacles.

Generally, in order to streamline the semiotic chaos that contradicts the sleekness of the official World Cup brand, FIFA constructs specific branded enclaves in host cities. Based on contracts[2] with the host (country, cities), FIFA controls and dictates the use of urban space, not only in the fenced FIFA Fan Fest zones (dedicated to screens, carnivalism and sales points of global brand products in the city centres and neighbourhood squares) and stadium interiors, but also within a one-kilometre radius of the World Cup stadiums, where people are allowed to sell only products of the official World Cup partners[3] on match days (Cornelissen, 2010, p.

---

[2] Initially, FIFA's sway over the World Cup hosts in the planning and branding of the urban space stemmed from the bidding process. In South Africa's victorious Bid Book, the Local Organising Committee and South African Football Association subscribed, for the purpose of persuading FIFA to award them with the right to host the mega-event, to a number of obligations. These include, among other things, tax-free rights around FIFA-designated sites and the protection of FIFA's intellectual property rights through the support of local police and other officials (Bond & Cottle, 2011, p. 47).

[3] In South Africa, the official Partners and Sponsors of FIFA were Adidas, Budweiser, Castrol, Coca-Cola, Continental AG, Emirates, Hyundai, Mahindra Satyam, McDonald's, MTN Group, Seara Alimentos, Sony, Visa and Yingli, whereas BP Africa, First National Bank, NeoAfrica, Passenger Rail Agency of South Africa, Shanduka-Aggreko and Telkom SA comprised the six National Supporters.

136). Also in South Africa, FIFA sued those infringing on the immaterial property rights of itself and its partners, such as the street vendors selling their homespun items. Legal proceedings have become a way to incorporate local symbols and concrete urban spaces "into the realm of the logic of product form" and thus enhance the economic and cultural power of transnational big business (Bruun et al., 2009, pp. 71-75).

When thinking about World Cup host cities in relation to media publicity, it is necessary to notice two closely related aspects that explain the enthusiasm for setting up specific enclaves in the cityscapes. Firstly, these FIFA-controlled "contractual spaces" (Augé, 1995, p. 101) are the stages on which the overwhelming majority of images and stories that circulate in the global media are played out. Also, tourists and locals mainly take their photographs in these spaces to memorialise their participation to the World Cup spectacle; when these end up on the platforms of social media, people's images largely distribute the same "emotion-rich iconography" as the television footage. To put it differently: the enclaves branded by official logos, slogans, colours and ideal fans are the channels and concentrations of the mass media (publicity) through which a harmonized tournament and city brand are communicated to the rest of the world.

Secondly, and more particularly in South Africa, while the media primarily celebrated the wow-architectural design of stadiums and the way people "lived the brand" in the enclaves, attention was shifted away from slum clearing and the increased intra-racial income and class differences in the South African rainbow society, as well as from the power politics and controversies over using public funds to build massive stadiums and the related new tourist-oriented infrastructure. Further, it is noteworthy that the above-mentioned enclaves were the locations where the urban regeneration projects were concentrated. In other words, redevelopment efforts were focused on specific "privileged event-led enclaves" (Cottle, 2011) rather than on the South African urban system or individual cities at large.

2. Fan performances as co-branding

At the event locations themselves, it was striking how the emotional intensity of the fans' behaviour always increased near the television companies' shooting crews. Groups of supporters were also actively gathered before the television cameras. As television companies in the

United States observed as early as in the Sixties: events should not be broadcast to people only because they are interested in them; people should be made interested in the events because they appeared on television (Powers, 1984).

Figure 1 - A television reporter interviews a group of fans of the Brazilian national team wearing carnival accessories. In the background, the Soccer City stadium (Photograph: Ilkka Immonen, 6/2010).

The fans' active involvement in creating the carnival mood, both at the stadiums and elsewhere in the urban space, was obviously in line with the image-building interests of the host, FIFA and the sponsors. The South African World Cup is a prime example of how a "hospitality society", striving for economic growth, may be defined, while organising a sports mega-event, and creating and presenting a "good vibe" as something of a national task (Whitson & Horne, 2006, p. 83). It was typical that before the tournament, through Football Friday happenings (in which people were asked to don national colours) and by a brewery-sponsored group known as the "super fans" touring throughout the country, people were advised on how they could achieve the "real" party mood, and they were encouraged to celebrate by wearing huge plastic spectacles and fanciful outfits–and if possible, to present themselves as a "celebratory bunch" before the television cameras.

In the ads, different types of "star commodities" (teams, players and products by specific sponsors) and fan images were offered as reference points for co-branding. The Open Happiness campaign, for instance, sponsored by Coca-Cola, capitalised on the legendary goal celebrations by Cameroon's Roger Milla and by other world-famous football players. Organised in the vicinity of stadiums and traffic nodes, like the Johannesburg airport, the campaign urged football fans to celebrate "openly" in the campaign booths reserved for this purpose. One of the most brilliant pictures that simultaneously built the tournament, country and city brand, and was widely circulated by the media, was of Nelson Mandela, ex-president and the epitome of the anti-apartheid struggle, posing with the World Cup trophy. Textual messages and stories worked powerfully with carefully selected images emphasising emotions. Through these, deliberately chosen facial expressions and body movements signalled the community-oriented and emotion-rich character of football.

On television, the atmosphere was created both by showing a panorama of the colour and sound landscape in the stand, and by cutting and zooming in to the spectators' gestures and facial expressions. By picking out football fans in their carnival outfits and splicing them into the stream of images, television acted as a kind of "educator in the correct way to feel", conveying how authentic and true fans of the sport behave and position themselves, not only in relation to the game itself and other fans, but specifically in relation to the camera. The presumption was that you had to please the camera by responding to its call, and enthusiastically gesturing and bellowing even if your favourite team was trailing. In the stands, there was an ongoing arms race among fans to be selected by the camera and thereby make it into global image circulation. Consumption, therefore, not only concerns person-object relations, but also person-person relations, whereby people observe and "consume" other people's acts and performances (Crawford, 2004, pp. 3-4). In all televised matches, celebrities of all kinds, from a wildly dancing Desmond Tutu to Mick Jagger, also acted as role models for "true" football enthusiasm (see also Lee, 2005).

Even though the national teams' flags and colours could be seen everywhere, the fan acts were not necessarily expressions of deep-seated (happy) nationalism (Manzenreiter, 2006, p. 146). As in other media-driven sports spectacles, the performances of the wildly celebrating visitors and the local fans can be seen as affirmative reactions to the co-branding invitation of the FIFA and the host, to being at the centre of global media publicity for a fleeting moment and to participating in producing the media spectacle of supporting (cf. King, 1997, p. 236; Kellner, 2003). The "cool" African fan spectacles at the tournament venues, and as portrayed in the media, were a strategic part of building the tournament, country and city brand. According to my observations, the celebration that took place in the official FIFA-controlled enclaves clearly lacked some of the signs of carnivalism highlighted as essential by Mikhail Bakhtin (1996), such as turning power hierarchies upside down, ridiculing the rulers and being accessible to all social classes and ethnic groups. In South Africa the tame fan acts rather contributed to endorsing the business and marketing strategies of officialdom, with the persuasion and exclusion desired by FIFA.

From the perspective of the host cities, the goal was that the affective carnivalistic fan utopia would give them a "semiotic head start" (Kainulainen, 2005, p. 108, p. 284) in the "place competition" for positive media publicity and the assumed cash flow resulting from that. In urban branding, which involves fan performances as its central strategy, the primary purpose is to flirt with some of the target groups put on a pedestal by contemporary capitalism: these notably include investors (mobile capital), tourists and taxpayers with good salaries (Hinch & Higham, 2004, p. 5). It was therefore important that people were at least momentarily made to believe that the consumption-based images were worthy of the new "world-class city" in order to sell places on the global market.

3. Transgressive fan acts and resistant activities

Undoubtedly, a large proportion of the fans responded positively to the call of marketers and television companies to act as an ideal fan–to participate in the (media) spectacle of support. Football fans are not, however, a uniform group; among them, you find divergent views on the sport's commodification, marketing and collaboration agreements with business and sponsors (see Giulianotti, 1999, 2002; Giulianotti & Robertson, 2009; Edensor & Millington, 2008; Kennedy & Kennedy, 2010). Therefore, I now shift my focus to the attempts of fans and activists to "do otherwise", through political demonstrations and to anti-brand meanings in the South Africa World Cup tournament. It is interesting how the mainstream media paid very little notice to the civic activists' voices of political resistance during the tournament: the attention was rather focused on some less politically motivated and non-organised transgressive acts by fans (for transgression and resistance in urban space see Cresswell, 1996).

One relatively innocent act that crossed the conventional boundaries of affirmative fandom was the exit of some of the South African (i.e., the home team) fans from the stands during the match against Uruguay, which the home team was losing. This scene was captured on television and was also analysed in the Finnish World Cup studios. The situation was not considered auspicious to the tournament's atmosphere. According to Martti Kuusela, a Finnish expert commentator, people's reactions showed that the local people had not quite bought into the event mood "heart and soul", after all. In South Africa, appeals by the representatives of the organising committee and celebrities were heard in various media, encouraging fans to continue sustaining a proper atmosphere as before. In re-engaging the fans, the spatial scale of fandom, and the scope of the imaginary community, were stretched more actively than before towards a pan-African fan identity. At the same time, the fans' spontaneous emotional reactions, which were labelled as wrong, were placed within the moralising framework of a "national task".

Another case that "disturbed" the "flawlessness" of the branded image and attracted a lot of attention in the mainstream media was the vuvuzela horn. This is a prime example of how football brands, tailored from various kinds of commercial and cultural elements, often include contradictions and meanings that do not appeal to everyone, despite all attempts at persuasion. Before the tournament, Hans Klaus, Director of Communications and Public Affairs at FIFA, said that banning vuvuzelas "would mean one would have to take away cow bells from Swiss fans and ban English fans from singing" (ESPN soccernet, 2009). However, during the tournament, many non-African visitors, players, journalists and television spectators, thought the sound landscape produced by the horns to be distracting and unworthy of football. Even though crossing the irritation threshold was problematic for the tournament, country and city brand, FIFA did not ban the use of the horns, as they were considered an organic part of African football culture. It was obvious that FIFA achieved wide sympathy by supporting the local fans in their battle for the right to define the football soundscape of Africa.

The case of Bavaria vs. FIFA was another scandal widely commented on in the media; the underlying cause is the strict ban by FIFA on ambush marketing at the World Cup stadiums by companies that are not official sponsors. The case was triggered by an episode in the Netherlands-Denmark match at the Soccer City Stadium in Johannesburg, in which a group of 36 female Dutch fans in skimpy orange dresses (i.e., the colour of their favourite team) were ejected from the stand by the organisers. The group was taken to the police for interrogation and threatened with prison sentences. The problem was that these (logoless) dresses had been handed out as free gifts with Bavarian beer packages earlier at the World Cup qualifying matches in the Netherlands. FIFA interpreted this to be an act of ambush marketing, whereby a Dutch brewery that was not one of its partners, tried to use the 'uniform sea of orange' and female beauty to capitalise on the media attention enjoyed by the tournament and to harm the interests of Budweiser, the tournament's official beer brand. In the eyes of the FIFA, which goes to extremes in terms of watching its business interests, the Dutch ladies in their wrong outfits were, as Mary Douglas (2000) put it, "dirt or substance in a wrong place"–in marketing lingo, a threat to sponsor visibility and streamlined branding. In another case, a local artist was selling T-shirts with a Fick Fufa slogan printed on them. By parodying the strict standardisation of the symbol environment advocated by the FIFA, the artist succeeded in challenging the sanctity of the FIFA brand and also in causing many people to at least momentarily reflect on their ethical values.

It is symptomatic that the political resistance related to South Africa's World Cup–protesting against inhumane material circumstances–that was more deliberate than the cases described above, attracted only little attention, at least in Finnish television broadcasts. Demonstrations only briefly reported, or not reported at all, included the numerous strikes before and during the tournament. At various tournament locations, security guards, for example, organised walkouts and demonstrations over lower than promised wages, but these were suppressed by the police. Instead of being interested in hearing and reporting that the profitability of the FIFA-led transnational corporation family that earned billions from the event was partly based on poorly paid service jobs, the mainstream media were more interested in the "African sounds" of the vuvuzelas that "disturbed" the match atmosphere and in the alleged infringement of intellectual property rights. The same observation applies to the demonstrations organised by civic activists in Johannesburg and many of the other host cities (including Cape Town, Port Elizabeth and Durban) that protested about the evictions and fear caused by the "cleaning projects" on World Cup sites, about the lower social classes' "right to the city" and about determining unofficial local business to be without legal protection (see also Desai & Vahed, 2010). As 20 Durban civic organisations bitterly protested in their joint statement published amidst the tournament ecstasy:

> Whereas the rich have benefited from the World Cup, the poor have not; the Zakumi doll mascot and other memorabilia were made in China not South Africa; Durban's informal street traders have been displaced and barred from selling in the vicinity of stadiums; and Durban fisherfolk have been evicted from the city's main North Beach and South Beach piers; and street kids were brutally displaced from central Durban in advance of the World Cup; […] FIFA's tourist initiatives are based on what it calls "luxurious ambiance" not working-class hospitality; promises of 450.000 international visitors for the World Cup were high overestimated (Pillay, 2010).

Although media attention at the World Cup mainly focused on "just" transgressive fan acts, instead of actual political protests, they nevertheless revealed the terms that were used to restrict the fans' actions. The host sought to gain an "image victory" by persuading and moralising fans under the auspices of a national task. FIFA, in turn, was hiding behind its brand rules, the control being based on sanctions supported by law. With the means described above, the fans' freedom to engage in a cultural (visible-making) process in urban space, and the chances of doing otherwise, were constricted. In this context, it is not surprising that some (fan) activists felt they were exploited, while the profits ended up in the pockets of the ultra-rich FIFA.

4. Conclusions: towards all-inclusive world-class city

When analysing urban branding through (mass) mediated branded enclaves and television images, it becomes obvious that football media spectacles have gained features which, instead of presenting the traditional attachments to the place, could be characterised as media-sexy topoporno. Henk van Houtum and Frank van Dam (2002), who have studied the commercialisation of football, use the term topoporno to refer to the mediated and commodified fan attachment that has emerged alongside, and even replaced, the earlier topophilic attachment: in topoporno, the attachment to place is ephemeral, and manifests itself in intensive images and acts constructed for the outsiders' gaze (see also Baudrillard, 1991, p. 12). In my view, the increasingly topopornographic football culture is about harnessing both the local community's sociocultural resources and the fans' involvement in a consumeristic and media publicity-serving way in order to capitalise on them for the purposes of branding (cf. Edensor & Millington, 2008, pp. 175-177; Bridgewater, 2010). This, in turn, serves FIFA's goals of maximising revenue from television and marketing rights.

For their part, topopornographic fanscapes complemented the "total approach" of tournament, country and city brand, defined prior to the event by FIFA and its allies. Brand consistency, as taught in the marketing textbooks, with the slogans, logos, identifying colours and stories all declaring the same positive message, was successfully made visible in South Africa (see Anholt, 2005; Kavaratzis & Ashworth, 2005). By focusing on the fans having a good time and on the correct way of displaying "sense of place" (Tuan, 1974; Bale, 1989), the material and social injustices related to the brandification of urban space were seemingly depoliticised, diverting attention from FIFA's rights secured by (emergency) law to control public space and to use the police when problems occurred.

It should be emphasised that in the context of football media spectacles urban branding is professionally systematic and socio-symbolically exclusive, and that, first and foremost, media visibility is a significant agent of change in urban spaces. According to critical voices, the South African World Cup paved the way for an increasing commercialisation, gentrification and socio-spatial polarisation of urban space (e.g. Pillay et al., 2009; Cottle, 2011). An issue in its own right is the fate of the many grandiose stadiums built for the World Cup that have ended up being underutilised since the tournament, with use and maintenance costs left for the local taxpayers to absorb. Standing almost empty, these numerous white elephants are landscape monuments that remind people not only of the past football fever, and global media attention, but also of the hugely oversized budgets and image-making hype prior to the World Cup (Kasimati, 2003).

As a real and permanent legacy–in the shadow of the manufactured national belonging and togetherness hyped by the mainstream media–there remain high levels of public and individual indebtedness, and reinforced social inequality in income, among and within cities. In a world-class city that is meant to be inclusive of all, investments should not be utilised just for one event; instead, the facilities built, or renovated, as a result of the investment, should be in efficient use, and available to all citizens, even after the event. There should be a primary emphasis on improving the existing infrastructure, especially in poorer residential areas, and not on building white elephants for scenic reasons, or on making political-ideological decisions to repair infrastructures located only on routes only favoured by tourists (Graham & Marvin, 2001). In addition, the "owners" of the media spectacle should contribute to bearing the financial risks by committing to paying part of the potential losses (Laakso et. al., 2006, pp. 96–97).

Presenting itself as a benevolent patron from one World Cup to another, FIFA keeps marginalising the local population for the sake of its own economic interests. One reason that FIFA's role in promoting neoliberal urbanism can remain invisible in the football media spectacle is because television journalists are not willing to "dig deeper". It would be interesting to see journalists take a more active role as a "mediators" between the speaker and the spectator: criticising, expounding on and explaining the different types of discourse to their audience (Ojajärvi & Valtonen, 2006, p. 195). This would minimise the possibility that speeches by those in power be authorised as public knowledge and the prevailing truth, and in a sense, as the voice of the entire nation (Nylynd, 2006, p. 158). This would also mean that urban branding, developed through football media spectacles, would present itself as more contradictory and multidimensional than could be concluded from the images and texts displayed by the event's production apparatus. It would be beneficial if journalists could perform deep and thorough analyses on who benefits from the new branded urban spaces, who is excluded from them, and what pockets of incorruptible resistance remain behind the topopornographic scene.

References

Anholt, S. (2005). Some Important Distinctions in Place Branding. *Place Branding*, 1, 116–121.

Augé, M. (1995). *Non-places. Introduction to an Anthropology of Supermodernity.* London: Verso.

Bakhtin, M. (1996). *Francois Rabelais – Keskiajan ja Renessanssin Nauru.* Helsinki: Taifuuni.

Bale, J. (1989). *Sports Geography.* London: E & FN Spon.

Baudrillard, J. (1991). *Ekstaasi ja Rivous.* Helsinki: Gaudeamus.

Bond, P., & Cottle, E. (2011). Economic Promises and Pitfalls of South Africa's World Cup. In E. Cottle (Ed.), *South Africa's World Cup. A Legacy for Whom?* (pp. 39–72). Scottsville: University of KwaZulu-Natal Press.

Bridgewater, S. (2010). *Football Brands.* Basingstoke: Palgrave MacMillan.

Bruun, O., Eskelinen, T., Kauppinen, I., & Kuusela, H. (2009). *Immateriaalitalous. Kapitalismin Uusin Muoto.* Helsinki: Gaudeamus.

Cornelissen S. (2010). Football's Tsars: Proprietorship, Corporatism and Politics in the 2010 FIFA World Cup. *Soccer & Society*, 11, 131–143.
Cottle, E. (Ed.). (2011). *South Africa's World Cup. A Legacy for Whom?* Scottsville: University of KwaZulu-Natal Press.
Crary, J. (2000). *Suspensions of Perception: Attention, Spectacle, and Modern Culture.* Cambridge: The MIT Press.
Crawford, G. (2004). *Consuming Sport:. Fans, Sport and Culture.* London: Routledge.
Cresswell, T. (1996). *In Place/Out of Place. Geography, Ideology and Transgression*. Minneapolis: University of Minnesota Press.
Debord, G. (1983). *The Society of Spectacle*. London: Rebel Press.
Desai, A., & Vahed, G. (2010). World Cup 2010: Africa's Turn or the Turn on Africa? *Soccer & Society,* 11, 154-167.
Douglas, M. (2000). *Puhtaus ja Vaara. Ritualistisen Rajanvedon Analyysi.* Tampere: Vastapaino.
Edensor, T., & Millington, S. (2008). "This is Our City": Branding Football and Local Embeddedness. *Global Networks,* 8, 172–193.
ESPN soccernet. (2009). FIFA give vuvuzela green light for World Cup. 17.7.2009, http://soccernet.espn.go.com/ news/story?id=661424&sec=worldcup2010&cc=5739.
Giulianotti, R. (1999). *Football. A Sociology of the Global Game.* Cambridge: Polity Press.
—. (2002). Supporters, Followers, Fans and Flaneurs. A Taxonomy of Spectator Identities in Football. *Journal of Sport & Social Issues*, 26, 25–46.
Giulianotti, R., & Robertson, R. (2009). *Globalization and Football.* London: Sage.
Graham, S., & Marvin, S. (2001). *Splintering Urbanism: Networked Infrastructures, Technological Mobilities and the Urban Condition.* London: Routledge.
Hinch, T., & Higham, J. (2004). *Sport Tourism Development.* Clevedon: Channel View Publications.
Hlatswayo, M., & Blake, M. (2011). Green Point Stadium: FIFA's Legacy of Unfair Play. In E. Cottle (Ed.), *South Africa's World Cup. A Legacy for Whom?* (pp. 225-254). Scottsville: University of KwaZulu-Natal Press.
Kainulainen, K. (2005). *Kunta ja Kulttuurin Talous. Tulkintoja Kulttuuripääoman ja Festivaalien Aluetaloudellisista Merkityksistä.* Tampere: Tampereen yliopistopaino.

Kasimati, E. (2003). Economic Aspects and the Summer Olympics: a Review of Related Research. *International Journal of Tourism Research,* 5, 433–444.

Kavaratzis, M., & Ashworth, G. J. (2005). City Branding: an Effective Assertion of Identity or a Transitory Marketing Trick? *Tijdschrift voor Economische en Sociale Geografie,* 96, 506–514.

Kennedy, P., & Kennedy, D. (2010). "It's the Little Details That Make Up Our Identity": Everton Supporters and Their Stadium Ballot Debate. *Soccer & Society,* 11, 553–572.

King, A. (1997). New Directors, Customers, and Fans: the Transformation of English Football in the 1990s. *Sociology of Sport Journal,* 14, 224–240.

Kellner, D. (2003). *Media Spectacle.* London: Routledge.

Kolamo, S., & Vuolteenaho, J. (2011). Kuningaspelin sisä- ja ulkopuoliset (Inclusions and exclusions in the world game: branded football spaces in South Africa's World Cup and England's Premier League.) *Alue ja Ympäristö*, 40, 17–34.

Laakso, S., Kilpeläinen, P., Kostiainen, E., & Susiluoto, I. (2006). Kisojen Aluetaloudelliset Vaikutukset. In N. Mats, S. Laakso & S. Ojajärvi (Eds.), *Urheilu, Maine ja Raha* (pp.73–98). Helsinki: Gaudeamus.

Lee, F. (2005). Spectacle and Fandom: Media Discourse in Two Soccer Events in Hong Kong. *Sociology of Sport Journal,* 22, 194–213.

Manzenreiter, W. (2006). Sport Spectacles, Uniformities and the Search for Identity in Late Modern Japan. In J. Horne & W. Manzenreiter (Eds.), *Sports Mega-Events: Social Scientific Analyses of a Global Phenomenon* (pp. 144–159). Oxford: Blackwell.

Massey, D. (2005). *For Space.* London: Sage.

Mayes, R. (2008). A Place in the Sun: the Politics of Place, Identity and Branding. *Place Branding & Public Diplomacy,* 4, 124–135.

Ngonyama, P. (2010). The 2010 FIFA World Cup: Critical Voices from Below. *Soccer & Society,* 11, 168–180.

Nylund, M. (2006). Suuri Lupaus: Kisojen Ennakointi Suomalaisessa Lehdistössä. In N. Mats, S. Laakso & S. Ojajärvi (Eds.), *Urheilu, Maine ja Raha. Tutkimuksia Vuoden 2005 Yleisurheilun MM-kisoista* (pp. 144-160). Helsinki: Gaudeamus.

Ojajärvi, S., & Valtonen, S. (2006). Kisat Mediajulkisuuden Puntarissa. In N. Mats, S. Laakso & S. Ojajärvi (Eds.), *Urheilu, Maine ja Raha. Tutkimuksia Vuoden 2005 Yleisurheilun MM-kisoista* (pp. 144-160). Helsinki: Gaudeamus.

Pillay, K. (2010). South Africa: World Cup for all! People before profit! Links: *International Journal of Socialist Renewal*, 17.6.2010, http://links.org.au/node/1747.

Pillay, U., Tomlinson, R., & Bass, O. (Eds.). (2009). *Development and Dreams: the Urban Legacy of the 2010 World Cup.* Cape Town: HSRC Press.

Powers, R. (1984). *Supertube: the Rise of Television Sports.* New York, Coward McCann.

Tuan, Yi-Fu. (1974). *Topophilia. A Study of Environmental Perception, Attitudes and Values.* New York: Columbia University Press.

Van Houtum, H., & Van Dam, F. (2002). Topophilia or Topoporno? Patriotic Place Attachment in International Football Derbies. *HAGAR, International Social Science Review*, 3, 231–248.

Whitson, D., & Horne, J. (2006). Underestimated Costs and Overestimated Benefits? Comparing the Outcomes of Sports Mega-Events in Canada and Japan. In J. Horne & W. Manzenreiter (Eds.), *Sports Mega-Events: Social Scientific Analyses of a Global Phenomenon* (pp. 73–89). Oxford: Blackwell.

Ward, S. (1998). *Selling Places. The Marketing and Promotion of Towns and Cities 1850-2000.* London: E & FN Spon.

CHAPTER IV

LOSING CENTRALITY: URBAN SPACES AND THE NETWORK SOCIETY

CHRISTIAN OGGOLDER

Introduction[•]

Cities are changing and they always have been. They have been growing or shrinking, they have been destroyed and rebuilt, they are melting into one another; in the end many are becoming what we today call mega-cities (e.g. Buijs et al., 2010; Sorensen & Okata, 2011). Although urbanisation continues, doubtlessly posing grave socio-economic problems, at the same time "dread continues that cities will disappear because of social, economic, political, religious, military, and natural forces, and changes in communication technologies–both personal and mediated". (Burd, 2007, p. 39)

My approach tries to combine those given facts using the notion of centrality and that of its loss. Thus, cities themselves will not disappear, but they are losing–or already have lost–their central position within networks–economic networks as well as communication networks. Moreover, urban changes, especially within the realm of communication, will continue as well. Places and spaces are moving into the virtual domain of the cities, and at the same time the virtual becomes real. People are increasingly accepting online activities as part of their real lives; the boundaries between two strictly separated spaces have been continuously blurring. In a way similar to media as *extensions of man* (McLuhan, 1964/2001) virtual spaces can be understood as extensions of cities. Based on an analysis of the historical development of urban and communication

[•] Many thanks to Katrin Kappler for her support and inspiring comments.

spaces, I aim to describe the integration of the urban space into the virtual space and vice versa.

1. Mutually Connected Nodes

Throughout history, cities always have been places of centralisation (Schäfers, 2010, p. 150). Cultural, political, and economic centres emerged from cultic centres. Mostly separated from their surroundings by massive city walls, protected by gates and with only a few distinct roads leading to neighbouring towns, the pre-modern cities formed clearly defined, mutually connected nodes within a net of connecting lines.

1.1. Some graph-theoretical remarks

Using the term "centrality", I refer to graph theory and social network analysis. Graphs within the framework of social network analysis describe a system of interrelated objects. These objects may be for instance persons, companies, or cities. In graph theory, a network is defined as a clearly specified set of *nodes* as well as the corresponding set of *lines* which connect the nodes (Oggolder, 2011, p. 335). According to this assumption, cities can be seen as nodes which are mutually connected by a variety of links. First and foremost these links are realised by streets and roads which form the basis of different kinds of exchanges between cities, usually persons, goods, or news. Depending on the particular importance of a town, the number of those interlocking roads varies. All roads lead to Rome, they say, and by this proverb nothing more than the outstanding centrality of the capital of the Roman Empire is expressed. I want to use ancient Rome as a prototype and an example of a city with a high degree of centralisation. Let us take a look at the basic road-system in ancient Rome (Fusero, 2009, p. 18).

As Figure 1 shows, the network of the Roman road system consisted of eight main roads which connected the most important cities of the empire with the capital. Most of the cities were directly linked with two neighbouring ones; therefore, the degree of centrality is two. Rome, however, was directly connected with five cities, which indicates a centrality degree of five. In terms of graph theory this means: "The more links a point holds, i.e. the greater the number of lines leading to it, the bigger its centrality is. Thus, the most central point within a network is that point, which has the biggest number of links." (Oggolder, 2011, p. 335) Not surprisingly, the capital held most of the direct connections to other cities; therefore, it was the most central city within the empire, and we can assume that this holds true for capitals in general. However, using raw centrality degree scores may lead to wrong conclusions because “the degree of a point depends on, among other things, the size of the graph, and so measures of local centrality cannot be compared when graphs differ significantly in size.” (Scott, 2000, p. 88) Thus, more meaningful than the absolute number of direct interlockings is the relative amount of local centrality (Freeman, 1979, p. 221) “in which the actual number of connections is related to the maximum number which it could sustain” (Scott, 2000, p. 88). Therefore, in our example Rome shows a relative centrality degree of 0.55 [= 5/10-1].

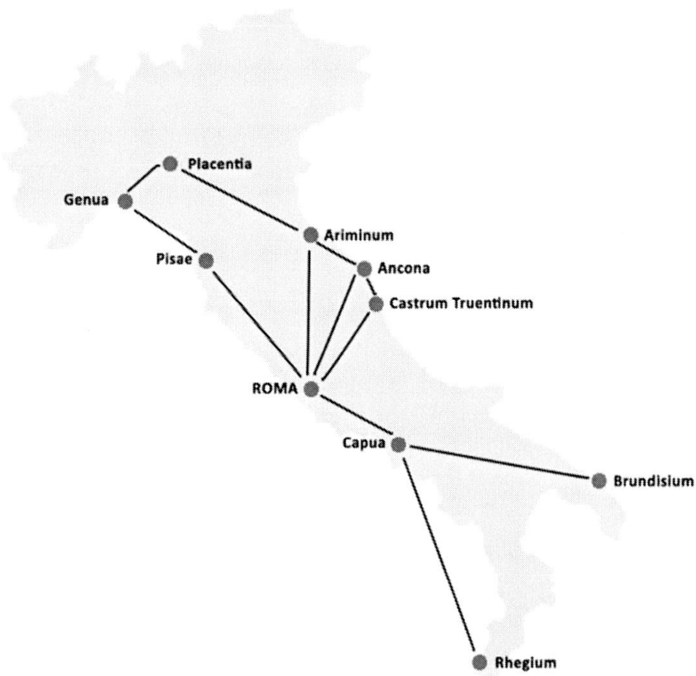

Figure 1 - Network of Roman roads

1.2. Moving closer

As a matter of fact, cities have always shown different degrees of centrality (Castells, 2009, p. 21), and in addition to that, I assume that the centrality of cities has been constantly *de*creasing in the course of time. This might seem an astonishing or even strange statement because what we all have doubtlessly experienced as a matter of fact is the enormous *in*crease of different kinds of links to and between cities. In the following I try to explain my assumptions.

The development of cities within one or more networks is characterised not only by a constant increase of their mutual connections but also by a development of different categories of interlocking. As I have argued before, the first and basic category of links between cities was systems of roads. As a consequence of technical innovations–mainly in the fields of transport and communication technologies–new layers of interurban exchange, new networks, have been established.

Besides constant improvements to road-systems, which enabled a more efficient exchange of persons, goods, and information, the invention of print–itself a new layer within communication networks–marks a qualitative advance in moving closer.

During the 16th century, printing–especially the illustrated broadsheet–developed from a so-called "special event" medium to an everyday medium. This becomes obvious if one takes a look at the various functions of typographic products (Schilling, 1990) in Early Modern history. Broadsheets and prints of that time have often been reduced by scholars to a propagandistic medium. As Michael Schilling says, broadsheets provided an essential part of everyday life and played a fundamental role in forming the early modern society (Schilling, 1990, 2).

The identical reproduction of information storages, a consequence of the invention of printing, has accelerated and broadened access to information. This led to more and more people becoming part of the new–what Michael Giesecke calls *the typographic data processing system*. Print not only changed media, it changed people's perception of reality (Giesecke, 1991, p. 21), and for that reason it changed spatial perception as well. Neighbouring towns moved closer via stories and news printed on circulating broadsheets, and therefore people could get an idea of even far away regions.

In addition to the new medium of print, the construction and development of communication lines between European centres during the 17th century, especially the establishment of permanent postal lines, is judged by Wolfgang Behringer (1999, p. 70) to be as important as the invention of print. Nevertheless, the system of a few central, mutually connected nodes still remained.

1.3. The claims of civility and nature

Printing not only changed people's perception of space in the sense of getting news from abroad as I have mentioned above. Given that *publish* means to send something to the *public*, the invention of printing also played a fundamental role with regard to the perceptual changes of public and public places. As Richard Sennett argues, "'public' thus came to mean a life passed outside the life of family and close friends" (Sennett, 1977, p. 17). When different domains of life became recognised as being defined for going public, others as a consequence became exclusively private (see also Jarvis, 2011, p. 69). "The line drawn between public and private was essentially one on which the claims of civility–epitomized by cosmopolitan, public behaviour–were balanced against the claims of nature–epitomized by the family." (Sennett, 1977, p. 18) The generation of a new sphere of communication realised by printing and commonly known as the public sphere has shifted real life, face-to-face communication within urban places into an abstract space, *linked* with the cities and *linking* to foreign cities as well. Following Colman and Ross (2010, p. 29), I argue "that the public has no ontological essence prior to mediated representation". These fundamental changes in the realm of public and public spaces on the one hand and privacy and family on the other have created constant preconditions for both modern media production and media consumption for centuries. Only within the last few years–as a consequence of social media of course–this normative framework is being abrogated. I will refer to this issue later.

Fundamental changes within this arrangement described above cannot be noticed until the Industrial Revolution. In addition to a restructuring of work and the creation of industrial centres, massive movements of people towards cities became a matter of fact. The result was an explosive growth of population in cities and the foundation of new towns. "Industry gradually made its way into the city in search of capital and capitalists, markets, and an abundant supply of low-cost labour." (Lefebvre, (1970/2003, p. 13) Since then, the conditions of existence have been linked to the city (see Löw, 2010, p. 605) and as a result of industrialisation, the city emerged as what Henri Lefebvre (1970/2003) has called "urban society".

On the macro level, the amount of links between cities and centres increased dramatically in order to fulfill these preconditions of modern life. New ways of economic production pushed the construction of more and better roads, railroads, and new communication technologies as well. On the micro level, however, the separate spheres of private and public fostered freedom as well as anonymity among citizens.

2. A New Arrangement of Space

2.1. Lost centrality

Today, a massive growth of metropolitan areas, which is increasing the numbers of street networks between major cities, is absorbing small centres, and the boundaries between urban and rural areas are becoming increasingly blurred (see Castells, 2001, p. 207). The development of cities and urban space has reached a stage absolutely contrary to the ancient and pre-modern town with clearly defined borders to the surroundings–ideally expressed by city walls. "So, the crisis is not just a matter of size", as Aurigi (2005, p. 10) states, but "it is also because the city is not a 'whole' any more or perhaps, it is not like the majority of us expect a 'whole' to be".

In addition to the breakup of clearly defined urban borders, today's cities are also enmeshed in a mostly uncountable number of networks. This enormous increase of inter- and intra-urban links has led to a considerable loss of centrality of cities within these networks. Referring to graph theory again, the degree of relative centrality not only depends on the number of links but also on the number of actors within a given network. So in the case of the above mentioned ancient Roman road system, if we add Ravenna for instance, directly connected with Ariminum, the relative centrality of Rome will decrease from 0.55 to 0.5, but at the same time, the centrality of Ariminum rises from 0.33 to 0.4. Depending on both the number of links and the number of nodes within a network, the number of indirect links rises, thus the centrality ultimately diminishes to lower degrees. As this very simple model shows, growing networks show a clear tendency to decentralise. Mega-cities, however, operating as hubs within global networks (Sassen, 2001), therefore, can be understood as an attempt at resisting the decentralising tendencies of networks and keeping alive former structures of centralised power and organisation.

2.2. Urban Spaces and the Network Society

As Manuel Castells points out, “the Internet is the technological basis for the organisational form of the Information Age: the network” (Castells, 2001, p. 1). As a consequence of the introduction of computer-based information and communication technologies, networks–being “primarily the preserve of private life” in contrast to “centralized hierarchies [which] were the fiefdoms of power and production”–have been enabled “to deploy their flexibility and adaptability, thus asserting their evolutionary nature” (Castells, 2001, p. 2).

According to the loss of its centrality, the urban space dissolves itself concurrently. “As urban history has shown, increased mobility–both physical and electronic–has reduced the need for spatial concentration”. (Carmona et al., 2010, p. 32) The virtual space of the web, however, has created a new arrangement of space. In addition to commonly used expressions of old and new media we therefore are confronted with spatial metaphors (Berker, 2001, p. 14) like *net* or *web*. We are invited to enter websites or to visit a company on the Internet; passwords prevent entrance to forbidden territories, crackers break into apparently protected rooms.

The image of two–originally–independent and self-contained territories is about to blur. Economy, politics, culture, and communication displace themselves from the urban space into the virtual space. Clearly defined borders of real and virtual rooms are vanishing, just like the definite boundary between centre and periphery. Via the Internet, which is more and more part of the real world, cities get the opportunity to establish new urban spaces. Below I will give three examples in order to amplify my assumptions.

The Tahrir Square in Cairo may serve as a vivid example for what I have just exposed theoretically. This square in the centre of the Egyptian capital was the location for demonstrations and protests long before the revolutionary movement of 2011. What was new in 2011, of course, was the use of new communication tools and social media. However, the Egyptian revolution was far from being an Internet revolution, contrary to what has often been asserted. The people demonstrated in real streets and did not lose virtual lives. Mubarak didn’t vanish by simply clicking the “I like” button. Accordingly, Linda Herrera (2011), for instance, argues on jadaliyya.com that it is high time to “insist that political and social movements belong to people and not to communication tools and technologies. Facebook, like cell phones, the Internet, and Twitter, do not have agency, a moral universe, and are not predisposed to any particular ideological or political orientation. They are what people make of them". There is nothing more to add. But what this event demonstrates perfectly is the integration of the virtual space into the urban. "Digital communications media have revolutionized learning, cognition, and sociability and facilitated the development of a new generational behaviour and consciousness." (Herrera, 2011)

Although studies have shown that the majority of Internet users act as consumers (Schmidt, 2009), we must not forget all those non-professionals who produce content which is usually summarised under the term "user-generated content". According to this, Axel Bruns identifies a considerable change in the balance between traditional media–such as newspapers, radio, TV–and the new players in the web–such as weblogs and news blogs. (Bruns, 2005, p. 2). Bruns describes traditional media as industrialised products, like things that we use in our everyday life. They are produced in an industrial, Fordist way, shaped as identical objects, and are homogeneous for each customer. In contrast, he characterises the new forms of news production as being "interactive and customizable by users, much in the same way that post-modern products frequently consist of a common central core, which can be modified and individualized through the addition of a range of accessories" (Bruns, 2005, p. 218). As a consequence, producers and consumers of news are no longer strictly separated. The individual–as a collective or a single person–is attributed more importance and has the opportunity to leave the traditional media production behind, shaping actively urban life and space. Today, the merit of *personalised media products* is increasingly recognised, and they are viewed as being akin to a handmade item. Thus, a new definition of journalism is required.

Finally, my third example refers to concerns often expressed about an increasing loss of social contact due to Internet use. As I have mentioned above, preoccupations about anonymity and social disintegration have been connected to urban life at least since the time of Industrialisation; moreover, Peter Burke (2006, pp. 26-30) has shown that these phenomena had been *topoi* throughout history. However, quoting Gordon and de Souza e Silva (2011, p. 86), "the web is brought into the spaces we occupy, and, similarly, those spaces are brought into the web. But this takes practice". Internet platforms of residential neighbourhoods provide an example for such practising. Web based communication technologies may serve as vehicles to overcome urban isolation and may foster interactions and social contact among residents, because "relationships are rarely maintained through computer-mediated communication alone, but are sustained through a combination of online and offline interactions" (cf. Hampton & Wellman, 2000, p. 207). Burd (2007, p. 40) claims a "new definition of the city" and in particular that the wireless world "mixes and melts the virtual and real, forces old media to merge and readjust, and even leads to the rediscovery and recovery of the interpersonal communication considered inherent in the historical reason for cities to exist".

3. Conclusion

Based on an analysis of the historical development of urban and communication spaces, the purpose of this paper has been to investigate the integration of the virtual space into the urban space. With an awareness of the spatial dimension of Internet communication, a re-theming of space concerning questions of mediatisation has to be considered in further studies (see Eckardt, 2011, p. 174). Following Veel (2006, p. 229) I argue that "also on a non-metaphoric level, the web and the city seem to be two increasingly intermingled entities: the organisation of the web draws from experience moulded by life in the modern city, and the city in turn depends on the infrastructural grid that the web adds to city life".

As a consequence, I assume that the fragmentation of society and its public sphere into segmented publics and individual interests must not be seen as the result of digital media. On the contrary, the newly established digital space of the Internet offers new places and rooms for gathering and political participation. The loss of centrality of both the cities and the owners of communication media opens up new opportunities for a less hierarchical and vertically organised but a more enhanced and horizontal communication landscape. Quoting Jeff Jarvis (2011, p. 59), the Internet "is a deep mine filled with unseen potential. It won't be used in all the ways we want. It will surprise us for good and bad. But it would be a mistake to declare one use or impact null just because an early encounter did not meet overblown expectations".

References

Aurigi, A. (2005). *Making the Digital City. The Early Shaping of Urban Internet Space*. Aldershot: Ashgate.

Behringer, W. (1999). *Veränderung der Raum-Zeit-Relation. Zur Bedeutung des Zeitungs- und Nachrichtenwesens während der Zeit des Dreißigjährigen Krieges*. In B. V. Krusenstjern & H. Medici, (Eds.), *Zwischen Alltag und Katastrophe. Der Dreißigjährige Krieg aus der Nähe* (pp. 39-82). Göttingen: Vandenhoeck & Ruprecht.

Berker, T. (2001). *Internetnutzung in den 90er Jahren. Wie ein junges Medium alltäglich wurde.* Frankfurt/M: Campus.

Bruns, A. (2005). *Gatewatching: Collaborative Online News Production*. New York: Peter Lang.

Buijs, S., Tan, W., & Tunas, D. (Eds.). (2010). *Megacities. Exploring a Sustainable Future*. Rotterdam: 010 Publishers.

Burd, G. (2007). *Mobility in Mediapolis. Will Cities Be Displaced, Replaced, or Disappear?* In S. Kleinman (Ed.), *Displacing Place. Mobile Communication in the Twenty-first Century* (p. 39-58). New York: Peter Lang.

Burk, P. (2006). *Imagining Identity in the Early Modern City*. In C. Emden, C. Keen, & D. Midgley (Eds.), *Imagining the City, Vol. 1. The Art of Urban Living* (pp. 23-38). Bern: Peter Lang.

Carmona, M., Tiesdell, S., Heath, T., & Oc, T. (2010). *Public Places Urban Spaces: The Dimensions of Urban Design*. London: Routledge.

Castells, M. (2001). *The Internet Galaxy: Reflections on the Internet, Business, and Society*. New York: Oxford University Press.

—. (2009). *Communication Power*. Oxford, New York: Oxford University Press.

Colman, S., & Ross, K. (2010). *The Media and the Public. "Them" and "Us" in Media Discourse*. Malden: Wiley-Blackwell.

Eckart, F. (2011). *Mediale Urbanität: Paradigmenwechsel von der europäischen zur medialen Urbanität*. In O. Frey, & F. Koch (Eds.), *Die Zukunft der Europäischen Stadt. Stadtpolitik, Stadtplanung und Stadtgesellschaft im Wandel* (pp. 173-188). Wiesbaden: VS Verlag für Sozialwissenschaften.

Freeman, L. C. (1979). Centrality in Social Networks: I. Conceptual Clarification. *Social Networks*, 1, 215-239.

Fusero, P. (2009). *E-City. Digital Networks and Cities of the Future*. Barcelona: List.

Giesecke, M. (1990). *Der Buchdruck in der frühen Neuzeit. Eine historische Fallstudie über die Durchsetzung neuer Informations- und Kommunikationstechnologien*. Frankfurt/M: Suhrkamp.

Gordon, E., & de Souza e Silva, A. (2011). *Net Locality: Why Location Matters in a Networked World.* Boston: Blackwell-Wiley.

Hampton, K. N., & Wellman, B. (2000). *Examining Community in the Digital Neighbourhood: Early Results from Canada's Wired Suburb.* In T. Ishida & K. Isbister (Eds.), *Digital Cities. Technologies, Experiences, and Future Perspectives* (pp. 194-208). Berlin: Springer.

Herrera, L. (2011). *Egypt's Revolution 2.0: The Facebook Factor.* http://www.jadaliyya.com/pages/index/612/egypts-revolution-2.0_the-facebook-factor (Download 14.01.12).

Jarvis, J. (2011). *Public Parts: How Sharing in the Digital Age Improves the Way We Work and Live.* New York et al.: Simon & Schuster.

Lefebvre, H. (1970/2003). *The urban revolution.* Minneapolis: University of Minnesota Press.

Löw, M. (2010). *Stadt- und Raumsoziologie.* In G. Kneer & M. Schroer (Eds.), *Handbuch Spezielle Soziologien* (pp. 605-622). Wiesbaden: VS Verlag für Sozialwissenschaften.

McLuhan, M. (1964/2001). *Understanding Media. The Extensions of Man.* London: Routledge.

Oggolder, C. (2011). *Graph theory.* In G. A. Barnett (Ed.), *Encyclopedia of Social Networks* (pp. 335-336). Thousand Oaks: Sage.

Sassen, S. (2001). *The global city: New York, London, Tokyo.* Princeton: Princeton University Press.

Schäfers, B. (2010). *Stadtsoziologie. Stadtentwicklung und Theorien – Grundlagen und Praxisfelder. 2., überarbeitete und aktualisierte Auflage.* Wiesbaden: VS Verlag für Sozialwissenschaften.

Schilling, M. (1990). *Bildpublizistik der frühen Neuzeit. Aufgaben und Leistungen des Illustrierten Flugblatts in Deutschland bis um 1700.* Tübingen: Niemeyer.

Schmidt, J. (2009). *Das neue Netz: Merkmale, Praktiken und Folgen des Web 2.0.* Konstanz: UVK.

Scott, J. (2000). *Social Network Analysis. A Handbook.* London: Sage.

Sennett, R. (1977). *The Fall of Public Man.* Cambridge: Cambridge University Press.

Sorensen, A., & Okata, J. (Eds.). (2011). *Megacities: Urban Form, Governance, and Sustainabilit.* Tokyo: Springer.

Veel, K. (2006). *CyberCitizen: Urban Identity in Net Art.* In C. Emden, C. Keen, & D. Midgley (Eds.), *Imagining the City, Vol. 1. The Art of Urban Living* (pp. 229-245). Bern et al.: Peter Lang.

CHAPTER V

IN THE MIDDLES OF URBAN SPACE: THE CASE OF CRITICAL CITY

FEDERICA TIMETO

Media do not coincide with technological objects. Rather, they comprehend experiences, representations, forms of relation, identity concepts and forms of social organisation which exceed the technological object *per se*. According to Raymond Williams (1974), we cannot isolate a technology from its context, nor can we abstractly consider the causes of media changes. Williams explicitly opposes his claim to Marshall McLuhan's position, which he considers as formalist and deterministic. According to Williams, McLuhan doesn't ever really consider media as social practices, but as "physical events in an abstract sensorium" (1974, p. 127), whose effects are the same for everyone, in the same time and space, without any consideration of the power relations which regulate their use and contents. Whereas the core of McLuhan's thought is usually associated with the definition of medium as message, the core of Williams' thought could be summarized in the formula "the content is the context".

What if we inverted the order of terms and superimposed the two formulas, so that the medium itself is seen as context, and abandoned the still modern form/content dichotomy? In this perspective, media would be the environments–a term which McLuhan himself uses beneath the more popular term "galaxy" introducing *The Gutenberg Galaxy* (1965)–where social forms emerge, an immersive common ground working even before some experiences and practices take place. The communicative "nature" of media, then, is not a complex of technological objects that influence a pre-existing society (or a society following them as an effect), but rather consists of the couplings and sociotechnical dynamics that emerge through communication and media practices. Thus, media are the environments where society and technology eventually co-emerge.

When talking about the circulation of information, it is usually taken for granted that information moves through space, but the opposite, i.e. that spatiality is constituted by information, is rarely considered. Elaborating on the concept of the total field awareness of contemporary media offered by Marshall McLuhan to move beyond the transmissive model of information, I propose to look at media of communication as *middles*, rather than as means, i.e. as mediating environments that perform connections and generate complex social relations among hybrid sociotechnical agents. Considering the "middling" aspect of media underlines their locative quality. Today, media of communication, whether relying on locative devices or not, work as locative media that address us and can be addressed in turn, enhancing a condition of social and spatial addressability.

In an essay entitled *Addressing Media*, J. W. T. Mitchell writes that "if media are middles, they are ever-elastic middles that expand to include what look at first like their outer boundaries. The medium does not lie between sender and receiver; it includes and constitutes them" (2008, p. 4). Mitchell's aim is to show how media should be addressed rather than simply understood, in order to, on the one hand, let the double communicative and spatial meaning of the term "addressing" emerge, on the other underlining the reciprocity of media as environments and of society as a complex system. Today I want to specifically focus on the idea of addressability that he, among other things, elaborates in this essay.

According to Mitchell, media should rather be addressed than understood, as McLuhan would put it, "as if they were environments where images live, or personas and avatars *that address us and can be addressed in turn*" (2008, p. 3, italics mine). Mitchell intends to expand McLuhan's idea of media, and at the same time criticise Williams' de-reification of media, which leads the latter to ignore the materiality of media in favour of social practice as the only locus where actions become meaningful. As a matter of fact, if social practices are mediated, this doesn't mean that they are media yet. The paradox of media, for Mitchell, lies in the connection, or we could say in the hybridization, of material aspects and social practices. Media are both the objects and the operations, both being a "medium through which messages are transmitted, and a medium in which forms and images appear" (2008, p. 8): in sum, they are both the complex of (human and non-human) technologies and the habitat where they *take* and *make* place (2008, p. 9). Even if, properly speaking, media do not have a specific address, they nonetheless address us, and in so doing they also acquire their spatio-temporal location (it must be once again noted that, in English, "to address" means both actions). Thus, for Mitchell, addressing media not only means encountering and challenging, but also locating them, and at the same time situating the analysis of media, which has to confront "its middling, muddling location in the midst of media" (2008, p. 18).

Sociology can deal with media of communication as tools, or means, that is as intermediaries transporting information, or as middles, that is as mediating environments that perform relations which in turn create complex social connections[1]. Following Mitchell, we see that media act as mediators in their addressing us and being addressed by us. Addressable media show that space cannot pre-exist its relations, but that contexts are continuously re-contextualized through them. Media perform new connections as active interfaces that mediate the sociospatial. This aspect is particularly relevant in locative media practices today: we can thus define locative media practices as those practices characterized by a participated relationality linking humans and machines through a diffuse addressability, in which the concepts of corporeality, materiality and location are performatively redefined. The performativity of locative media practices at the same time appears as an embodied experience of the body/mind ensemble in a media environment which is more and more immersive and localized, and as the mapping of a material/informational spatiality which is itself performative.

Such a mobilization of the "where" at the same time intensifies and redefines *location a-whereness* (Thrift, 2008, p. 166) rather than erase it, contrary to how it might appear. Actually, considering space and mobility together has a double consequence. On the one hand mobility, be it material or not, cannot be seen as a passage from one point to another in space, but as the possibility of producing and consuming information in movement. On the other hand, the "where", mobilized by information, disengages spatiality from a purely dimensional perspective, linking it to the practices of the everyday. This underlines a reciprocal co-emergence of codes and socio-spatial formations, pointing to the performativity of both.

---

[1] I am referring, here, to the distinction between the notion of *intermediary* and that of *mediator* proposed by Bruno Latour (2005). Shifting the focus of attention from media as intermediaries to media as mediators means taking into consideration the way the forces of change take place and combine in a contingent and heterogeneous way, rather than the causes determining social change.

As the space of a distributed materiality (The Internet of Things) and distributed information (Ubiquitous Computing), such a space continuously happens, relating subjects, objects and places in everyday practices. In order to see how, I want to share my preliminary thoughts on the game Critical City Upload (CCU). This case study is part of my doctoral thesis at the University of Urbino, in which I focus on locative media and location-based social networks (LBSNs), in order to investigate the socio-spatial relations taking place in so-called *mediaspaces* or *codespaces* (Couldry & McCarthy, 2004; Kitchin & Dodge, 2011). In particular I consider CCU as an example of how media can be locative even when they do not strictly rely either on a digital infrastructure or on mobile devices[2]. More specifically, I am interested in studying how space traversed and mapped through locative media practices is both the theatre and the product of distributed agency.

Given that my research question asks what the implications of this renewed location awareness of media are, the CCU case specifically allows me to verify three of my six operative hypotheses, that is: 1) how and to what extent do locative media and applications work as middles for social actors in a condition of reciprocal addressability? 2) Can we speak of a new emerging "participative urbanism" (Paulos, Honicky & Hooker, 2009) related to the diffusion of locative media? 3) How are locality and location performed through locative media, once proximity is considered not only a physical, "immediate" issue but also an informational one (either digital or not)?

In the words of its developers, CCU is a non-profit "collective game of urban transformation", which also works as a social network. Conceived in Italy and launched as a desktop application in 2008, CCU (whose second season started in November 2011) aims to alter the perception and experience of public space and to offer prototypes for real interventions. Although, strictly speaking, CCU does not presuppose the use of mobile devices, the game can be said to be locative for several reasons. The positions of participants and their performances are geolocalized; the game comprises a set of instructions divided in different levels, which can be performed by going "outside" and documented by means of texts, pictures and videos, implying different forms of mobility and mobile tools; finally, there are "nodes", defined as "special places where strange things

[2] See Gordon & de Souza e Silva (2011); see also the very interesting article *An Internet of Things* by Keller Easterling (2012), where the author discusses the fact that space can be read informationally independently from digital platforms.

happen", which are existing zones with public access where the accomplishment of a mission is more convenient, being associated with a higher score (attributed by the CCU staff). Every performance must be exactly geo-tagged according to the rules of CCU (which means that, when the missions are on the move, the most significant place where they happen must be indicated).

When at least 3 missions take place within 50 square meters, the staff verifies that the instructions are correctly followed and successfully performed (that is, accomplishing the rules of the game) and validates the node. Special nodes can also be inaugurated even if less than 3 missions have been accomplished there[3]. A node is composed of a geographical name (of the city) plus the geographical coordinates, an identification number, and a significant name or address further specifying its location. Each node is also given a bonus score–Gold (100 points), Silver (50 points), Bronze (25 points)–which means that from the moment a node is published on the home page of CCU, every mission performed in that node gets extra bonus points. Missions can be accomplished individually, in groups and also with the help of self-offering "collaborators", an aspect which reinforces the practices of microcoordination and socio-spatial networking in the mediated urban environment. For all these reasons, CCU foregrounds most of the aspects of the addressability of contemporary media of communication, offering an interesting example of the performativity of socio-spatial networks in the middles of information and communication.

In my research I adopted an ethnographic approach including participatory observation in order to study the genesis, characteristics and practices (comprehending both the "performances" of the chosen "instructions" and the online conversations commenting them, when available) of the CCU nodes, which are 51 so far. However, for reasons of time, today I am only able to talk about one node, which is currently the first as it has the largest number of players involved, because I believe that it exemplifies very well the socio-spatial dynamics at work in this urban game. Since I am only at the beginning of my analysis, I do not have any conclusions yet, but I want to share my initial observations with you.

---

[3] These usually are places where private or public organizations pursuing activities oriented towards the improvement of the local reality work.

The node I talk about is Trento 02 Piazza Fiera, a silver node that counted 25 players for 10 accomplished missions at February 2012 (all of them belonging to the 2011 season of the game). *Andate e Moltiplicatevi* realized by *Logan* was the first mission accomplished in the node: the instructions required three flyers about CCU to be attached at three different bus stops. It is from this apparently very simple mission that the community around the node originated. Actually, one player, *The Small*, affirmed many times that he started playing after seeing these flyers; significantly, he also accomplished a mission in this node, entitled *A nuova vita* that, while requiring that something broken or needing some adjustment in the street is fixed by the player, in this case also relates to the origin of the node. In fact, *The Small* decided to reinstall the flyers of the first mission realized by *Logan*, which were no longer in their original position (instead of placing his own ones created according to the same instructions). In so doing, he recreates the original piece of spatial information, while adding a further layer to it, one comprising the story of his entry into the game as well as of his friendship with *Logan*[4].

Additionally, if we look at the definition of "participatory urbanism" by Paulos et al. (2009), this mission can be said to have generated "individual and collective needs based dialogue tools around the desired usage of urban […] spaces" (p. 420). The relational potential of a place like the bus stop, where many people usually stand together alone waiting for the bus, was suddenly put forth by a very simple communicative action relying on a basic tool, a printed piece of paper. And, as we will see, this is the same dynamic of many other missions accomplished here, particularly those realized in group and with the involvement of local people, like *Ampie strette* e *Le forme dell'aforisma.*

If the scholarship about OSNs generally underlines how social network users tend to reinforce existing offline relations, we can conversely notice how in locative social networks like CCU we also have the opposite tendency; social relations that are initiated online are cultivated in the physical context of the game performances over time. This by the way confirms the permeability and continuity of the two dimensions, while also highlighting the hybridity of (net-) locality, in which "co-presence is not mutually opposed to networked interaction" (Gordon & de Souza e Silva, 2011, p. 87).

---

[4] Let us note that the mission is accomplished according to minimal logic very similar to that of geocaching.

Since the missions in this node relate to each other, they are like chapters of a single narration linking the online and the offline environment by means of texts (both descriptions and conversations) and images. People living in the same city get to know each other through the game, meeting in a common space represented by the node which is an actual geographical location as well as a space of material and symbolic performativity. As Gordon & de Souza e Silva (2011) put it so well, it is the circulation of communication that creates the awareness of an existing community as well as the concrete possibility of connecting at a local level. Performances do not simply express some specific aspects of the node, rather they disseminate information creating communicative links that relate people who already know each other to unknown people.

If we follow the story of the node players, we can also interestingly note that some of them also play in other nodes or even contribute to their creation not necessarily in the same geographical area, occasionally also bringing their playmates with them (this is for example the case of *Logan*, who was from Rovereto but went to Rome where his friend *Frida* helped him create the Roma Colosseo node). Another interesting aspect is that the conversations developing in the CCU social network and in the related FB page not only presented or commented on the performances effected, but also acted as spaces to meet and plan the next ones. Thus, CCU works as a middle composed of more or less intense nodes whose augmented spatiality recreates a dimension of information and communication each time preceding but also exceeding the territorial locations, bringing together different forms of physical and digital proximity.

Whereas the core group of players for each node remained more or less the same across different missions (confirming that the creation of a node is a collaborative and not an individual effort), some missions, like *Ampie strette* (significantly retrieved in the second edition by The Small) or *Le forme dell'aforisma*, involve local unknown people (or even people 'made local' for the occasion, like in *Le forme*, where players who come from out of town for the mission are significantly "baptized" to properly take part in the performance). *Ampie strette*, for example, conceived as a ludic flash mob, required at least 30 unknown people to be hugged on the street.

The mission was accomplished during the Italian Carnevale, a moment when the transgression of the everyday routine is usually more accepted, by a group of ten players, and it was interesting to observe both the description and the comments following the publication of the mission online. In fact, starting from the choice of the moment of the year to perform the instruction, all the players' words underline the pleasure and fun of involving other people in a free and essentially relational activity such as hugging. Even if most of them note the initial difficulty of the performance (touching someone you don't know, but also doing it in a territory considered "cold" in the national imaginary), they underline the emotional involvement and the satisfaction for its final success, privileging the importance of the action as such over its documentation. I consider this an example of how location is, to borrow Mitchell's words, a "middling and muddling" environment rather than being a static background, so that sometimes it becomes very difficult to separate media forms from media contexts, documentations from performances.

One last concluding observation regarding the game logic of CCU, which I only indirectly focused on: the ephemeral and often minimal aspect of many of the missions of CCU, at least those accomplished in the examined node, underlines the participatory logic of CCU also beyond the content of the single missions. As again Gordon and de Souza e Silva note, the locative game brings people to explore locations, but it is "the resulting social encounters that enhance location awareness" in the end (2011, p. 68).

References

Couldry, N., & McCarthy, A. (2004). *Introduction.* In N. Couldry & A. McCarthy (Eds.), *Mediaspace: Place, Scale and Culture in a Media Age* (pp. 1-18). London, New York: Routledge.

Easterling, K. (2012). An Internet of Things. *e-flux*, *31*(1). Retrieved from http://www.e-flux.com/journal/an-internet-of-things/

Gordon, E., & de Souza e Silva, A. (2011). *Net Locality. Why Location Matters in a Networked World.* Malden, MA: Wiley-Blackwell.

Kitchin, R., & Dodge, M. (2011). *Code/Space. Software and Everyday Life*. Cambridge, MA: The MIT Press.

Latour, B. (2005). *Reassembling the social. An Introduction to Actor-Network-Theory*. New York: Oxford University Press.

Mcluhan, M. (1965). *The Gutenberg Galaxy. The Making of Typographic Man.* Toronto: University of Toronto Press.

Mitchell, J. W. T. (2008). Addressing Media. *Media Tropes eJournal*, *1*, 1-18.

Paulos, E., Honicky, R. & Hooker, B. (2009). *Citizen Science: Enabling Participatory Urbanism*. In M. Foth (Ed.), *Handbook of Research on*

Urban Informatics: The Practice and Promise of the Real-Time City (pp. 414-437). New York, London: IGI Global.
Thrift, N. (2008). *Non-Representational Theory. Space/ Politics/ Affect.* London, New York: Routledge.
Williams, R. (1974). *Television, Technology and Cultural Form.* London: Collins.

CHAPTER VI

SPACE AND THE GEOGRAPHICAL IMAGINATION ON THE DUBLIN DOCKLANDS

MOIRA SWEENEY

Introduction

> *The history of the Dublin Docklands can be read as a microcosm of the history of Ireland, both ancient and modern. The making of the Docklands is a complex tapestry of great engineering achievement, visionary planning, intrigue, economic rise and decline, and human triumph over adversity.* (Dublin Dockland Development Authority, 1997, p. 18 cited in Moore, 2010, p. 15).

The Dublin Docklands are of course more than a geographical space; this is a multi-dimensional space, a product of many forces including historical and economic necessity, business elitism, globalisation, contemporaneous regeneration and cultural affiliation. The above statement by the DDDA acknowledges the complexity of the strands that have been interlaced over centuries in the construction of the Docklands. In painting a polished image of this achievement as heroic it may obscure the more troubled social and economic histories of local indigenous communities. According to Held, globalisation can be understood as "spatiotemporal processes of change which underpin a transformation in the organisation of human affairs by linking together and expanding human activity across regions and continents" (1999, p. 15, cited in Moore, 2004, p. 213). Whilst it is not within the scope of this paper to analyse in depth Held and his colleagues' studies on globalisation, it is now recognised that any concept of globalisation as fundamentally geographical may overlook a more disruptive conjuncture of narratives. As Massey poses:

> Clearly the world is not totally globalised [...] as in the case of modernity, this is geographical imagination which ignores the structured divides, the necessary ruptures and inequalities, the exclusions, on which the successful prosecution of the project itself depends. [...] This is again–a geographical imagination which ignores its own spatiality (2005, p. 84).

Massey's inspired arguments for rejuvenating our imagination of space, globalisation and place are well established. I wish however to investigate with fresh eyes how they perform in a space such as the Dublin Docklands.

1. Beyond Geographical Space

A re-orientation of space invites us to consider it as "not a mere surface, or a single narrative but a multiplicity of trajectories" (2005, p. 9). Massey boils down her argument on space into three key propositions, proposing firstly that we recognize "space as the product of interrelations; as constituted through interactions, from the immensity of the global to the intimately tiny" (2005, p. 9). The Dublin Docklands is not a homogenous site but a space that has emerged as a consequence of a complexity of connections at play. It is the very interaction of historical and contemporaneous forces from the "immensity of the global to the intimately tiny" that makes the Dublin Docklands a socially, physically and economically multi-dimensional space.

Dublin Port in a small harbour on the River Liffey was initially within the confines of the city, but it expanded in the nineteenth century and was relocated eastwards away from the centre and close to the sea (Moore, 2010, p. 12). This strategically positioned port then emerged into a highly advantageous space for trading ships and hence Ireland's economy. Residential port workers' communities developed along the north and south quays with most working men employed as labourers and carters on the docks and women as domestic workers in other parts of the city. This docklands space already in its early formation was the product of the interrelations of geography, economics and the social as constituted through the interaction of the immensity of the global (trading ships) and the intimately tiny (docker communities). But this was not and is not a fixed space.

Massey secondly proposes that we recognize "space as always under construction [...] always in the process of being made" (2005, p. 9). The dockland communities were set east from the centre of the city, whether in

Ringsend or Sherriff Street and they were considered to be on "the wrong side of the tracks", literally meaning that they were the other side of the main railway line (Brady & Simms in Moore, 2010, p. 12). By the twentieth century the area became home to the poorer strata of society with insufficient amenities or recreational facilities and a lack of open space (Moore, 2004). The post-World War II international growth of container traffic and shift from rail to road led to loss of employment on the Dublin docks. This gave rise to rapid physical degeneration, poverty and intolerable housing conditions considered at the time reminiscent of Third World cities (ibid.). By the 1980s, the rupturing of the traditionally tight-knit communities resulted in the emergence of serious socio-economic difficulties, social disorder, vandalism and the area yielded to a flood of heroin and drug dealing as an alternate way of life. Caught in a particular moment of time, one image of this space in the mid 1980s could have shown widespread poverty co-existing with international trading. Continuing global and local change, however, over the next three decades confound this image of the docklands' space.

This leads to Massey's third proposal that we understand space as the "sphere of possibility of the existence of multiplicity in the sense of contemporaneous plurality; as the sphere in which distinct trajectories coexist; as the sphere of coexisting heterogeneity" (2005, p. 9). Certainly, the docklands' space offered and continues to offer a sphere of possibility, but the distinct trajectories that have emerged over the late twentieth century have not had an easy coexistence. Global technological and economic changes resulted in increased mechanisation and labour decasualisation on Dublin's docklands. Whilst the so called "new world" practices as performed on the docks were considered progress for, amongst others, the shipping companies, since the 1980s they have been a devastating blow for former dockers. For stevedore companies to survive the demands of a once militant workforce, labourers were disciplined and controlled with lower wages whilst automation lessened employment opportunities (Sekula, 1995). The residential docking communities' plight was further compounded by migration from the area with the neighbourhood profile becoming one of vulnerable and largely elderly, unemployed and educationally disadvantaged people (Moore, 1999).

A dramatic and unexpected trajectory then came into play, when, after a period of retreat from the waterfront, the docklands became a target for development renewal projects (Kokot, 2009). As part of an urban renewal programme in the late 1990s, the newly named Dublin Docklands became a "hotspot" or "engine of the Celtic tiger boom" (Moore, 2005, p. 219). Land prices soared and the increase of local housing prices introduced a new urban gentry which excluded the remaining indigenous community; the houses were out of the affordable house bracket for those who had grown up in the area resulting in many having to move out. The reconstruction of the Docklands by a business elite may have resulted in rich and poor living in geographical proximity but with a huge socio-economic gulf between the two. The economic restructuring of the last two decades has re-shaped this society and space resulting in further socio-economic exclusion, polarisation and invisibility. The local communities use the term "economic cleansing" to describe what has happened in their neighbourhoods (Hogan, 2006, p. 32). Rather than the distinct trajectories of business elitism and social need coexisting, "formerly cohesive neighbourhood identity and class structure has been fragmented" (Hogan, 2005a, p. 121). Urban regeneration, gentrification, globalisation, business elitism and transnational shipping have irreversibly impacted the traditional residential and working communities along the Dublin Docklands (Hogan, 2006; MacDonald, 2000). The space that is the Dublin Docklands has now transformed beyond recognition. As Massey argues:

> But places go on without you. A nostalgia, or a set of expectations that does not take account of that deprives others of their agency, denies their ongoing histories. It converts their coeval, different space into a moment in your time. In a move that is a form of colonisation, it holds others still (2003, p. 115).

A study of the transformed space of the Dublin Docklands is therefore only complete when the narratives and histories of those rendered invisible are woven into the "complex tapestry of great engineering achievement, visionary planning, intrigue, economic rise and decline, and human triumph over adversity" (DDDA, 1997, p. 18).

2. Space and the Senses

I argue furthermore that a revitalised geographic imagination must also extend to include the sensuous nature of Dublin Port; images, sounds, smells. Sekula however signals the disappearance of these elements:

> In the past, harbor residents were deluded by their senses into thinking that a global economy could be seen and heard and smelled. The wealth of nations would slide by in the channel. One learned a biased national

physiognomy of vessels; Norwegian ships are neat and Greek ships are grimy. Things are more confused now (Sekula, 1995, p. 12).

Until very recently Dublin's docklands were primarily viewed as a maritime district beyond the city centre. Being an island, Ireland's key means of trading has traditionally been through her seaports, with Dublin Port handling over two-thirds of containerized trade to and from Ireland and 50% of all Ireland's imports and exports (Dublin Port Authority, 2010). After almost two decades of the Celtic Tiger[28], the ugly urban scenes of the 1980s along the docks have been replaced by the sheen of post-modern architecture and engineering. Economic internationalisation and rejuvenation of the docklands have resulted in Dublin Port becoming part of "a fluctuating web of connections between metropolitan regions and exploitable peripheries" (Sekula, 1995, p. 48). Globalisation and information technology have therefore contributed to a new spatiality which centres on cross-border connections (Sassen, 2000). The activity on the Dublin docks which operates as part of this complex global digitalised structure is largely invisible to those working and living within a stone's throw from the port. Cheap goods from South East Asia can be in Dublin within a month. Moreover the containerisation of cargo movement, pioneered by the US shipping companies in the 1950s, has reduced loading and unloading time. Thus there are greatly increased cargo loads along the docklands but we don't know what they are. There are no longer smells or sights, just sanitised containers. As Sekula points out "despite increasing international mercantile dependence on ocean transport, and despite advances in oceanography and marine biology, the sea is in many respects less comprehensible to today's elites than it was before 1945, in the nineteenth century, or even during the Enlightenment" (1995, p. 4). By contrast, oral history, according to Kearns, recalls the Dublin docks as "a world of masts, funnels, towering cranes, barges, carts, horses… a hundred sounds becoming a symphony of dockland" (1991, p. 30). Archival film and photographic imagery from the 1950s[29] depicts cattle movement across the mouth of the River Liffey onto the docks. Today along the docks, on first sight, there is little sense of what goods are being moved in containers and, moreover, there is little to indicate the importance of the

---

[28] The Celtic Tiger is a metaphor coined by US Investment Bank Morgan Stanley in August 1994 and has become an accepted term for the rapid growth and transformation of the Irish economy in the 1990s.

[29] This material can be accessed in the RTÉ and Irish Film Institute archives and Alan Martin's private archive on www.bluemelon.com as well as the Dublin Dockworkers Preservation Society's collection.

role that the sea plays in transporting crucial commodities to and from Ireland. However, if you move beyond the scenic views of the Dublin Docklands, and venture deep into the heart of the South Coal Quay, you might come across the grime and dust of dockworkers unloading vital commodities such as eco cement or coke which have travelled from Russia on a Greek-owned, Polish-captained ship.

When Friedrich Engels set out to describe the living and working conditions of the English working class, he began by standing on the deck of a ship in 1844. He describes moving down the river of the Thames from the open space of the sea: "The further one goes up the river the thicker the concentration of ships lying at anchor. All this is so magnificent and impressive that one is lost in admiration" (Engels, 1845, p. 30 in Sekula, 1995, p. 42). This wonder subsides as he moves from the panoramic space of the sea to an "ugly" (1995, p. 30) urban scene in the closed slum spaces of London's main streets. Perhaps if he had travelled from Dublin Bay down the mouth of the Liffey less than thirty years ago his narrative shift from a magnificent panoramic maritime space to the "brutish frictions" (1995, p. 30) of urban life may have been very similar to that of 1845 London.

Sekula urges us to turn our consciousness back to the sea, the forgotten space, and recognise the importance of *maritime space* as opposed to the persistent focusing on cyberspace and the illusion of an instantaneous connection between far-flung lands (1995). The concept of maritime space has seized and fired my imagination for some time, inspiring the creation of an audiovisual representation of the complexities of transnational shipping in Dublin Bay and deep into Dublin Port. For this three-screen installation, *Sensing the Local* (2010), I filmed along the docks and from one of the Dublin Port Company boats with the help of the men charged with checking the buoys that guide ships in and out of Dublin Bay. For one of the screens, I situated a camera on the front of the boat and began by filming out towards open sea as if from the point of view of the boat, moving through uninterrupted space and the wide expanse of Dublin bay. I wanted to re-create a sense of the panoramic boundarylessness that has so occupied the imaginations of artists from Turner to Sekula. For the second screen, I allow cargo, passenger and container ships to cross the field of vision as they arrive into and depart from the docklands. For the last screen, I observe cargo movement and docker activity on the quays deep into the heart of the docklands.

SEASHARK

Figure 1 - *Sensing the Local*, 2010, Video, Screens I, II, III.

Over time, the three screen installation evolved into a multi-media installation 'Stevedoring Stories' which was exhibited as part of Dublin Tall Ships 2012. The impulse behind both artworks was to depict the multi-dimensional nature of contemporary maritime space: the ship, the dockland, the local working community, the visiting crews, the globally linked technology, the containerisation of cargo, the hinterland, the developed shorelines, and so on. In contrast to Engels' disappointment in the mid-nineteenth Century, in the 21st Century, I found that in moving from the panoramic expanse of Dublin Bay into the heart of the port, I was able to construct a contemporary working "symphony of dockland". This was my own Foucaultian "heterotopia" (Foucault, 1986, p. 24); a real space with the sounds of birds, foghorns, cranes and moving trucks singing to all the senses.

References

Dublin Corporation. (1986). The Inner City Draft Review.

DDDA. (1997). Dublin Docklands Development Authority master plan.

Engels, F. (1845). *The Condition of the Working Class in England.* Trans. and ed. by Henderson, W. O. & Chaloner, W. H. (1968). Stanford: California University Press.

Foucault, M. (1986). Of Other Spaces. *Diacritics*, *16* (1), 22-27.

Held, D. (1999). *Political Theory and the Modern State.* Cambridge: Polity Press.
Held, D., & McGrew, A. (2007). *Globalization/Anti-Globalization, Beyond the Great Divide.* Cambridge: Polity Press.
Hogan, J. P. (2005). Challenge, Renegotiation and Change in the Current Phase of Spencer Dock. *Progress in Irish Urban Studies*, *1*(2), 13-20.
—. (2006). The Politics of Urban Regeneration. *Progress in Irish Urban Studies*, *2*, 27-37.
Kearnes, K. C. (1991). *Dublin Street life and Lore: An Oral History.* Dublin: Glendale Press.
Moore, N. (1999). Rejuvenating Docklands: The Irish Context. *Irish Geography*, *32*(2), 135-149.
—. (2008). *Dublin Docklands Reinvented.* Dublin: Four Courts Press.
McDonald, F. (2000). *The Construction of Dublin.* Kinsale, Ireland: Gandon.
Massey, D. (2003). *Some Times of Space.* In S. May (Ed.), *Olafur Eliasson: The Weather Project* (pp. 107-118). London: Tate Publishing.
—. (2005). *For space*. London: Sage.
Sekula, A. (1995). *Fish story.* Dusseldorf: Richter Verlag.
Wright, T. (1997). *Out of Place: Homeless Mobilizations, Sub cities and Contested Landscapes.* New York: State University of New York Press.

PART II.

MEDIA PRACTICES IN URBAN CONTEXTS

CHAPTER VII

THE SENSE OF PLACE FROM MOBILE COMMUNICATION TO LOCATIVE MEDIA

BARBARA SCIFO

Introduction

The paper intends to discuss the experience of place related to the use of location-based social networking through mobile applications (in particular Facebook Places), taking into account previous forms of social appropriation of mobile communication.

I will refer to some empirical findings–collected during different phases of the social history of mobile technologies–to show different relations and connections created between the subjects' physical space, connected with a specific social situation, and the virtual space of the communication mediated by mobile and locative technologies. In this way I hope to highlight the strong continuity between the social uses of mobile communication and localization services, rather than the novelty and socio-cultural discontinuity.

First let me briefly outline the theoretical frame–that is those theories, now largely credited in media studies (see Couldry & McCarthy, 2003), refuting Meyrowitz's idea (1985) that media's alteration of space restrictions causes a loss of sense of place as physical and symbolical dimension, significant for the definition of the situation in social interaction. These approaches also reject Castell's idea (1996) that space and place are absorbed and nullified in the space of flows of ICTs.

In opposition to the rhetoric of "death of geography", of the "contracting world" (Harvey, 1989) and that places are "dissolved" into "no places", these approaches underline the located feature of the ICTs as well as the need to look at the social situated uses of the new media (as the domestication paradigm suggests, see Silverstone et. al., 1992). So, in this theoretical context, special attention is given to processes of re-territorialisation and re-embedding of social relations extended in space, also thanks to media, in this way fostering mobility and networking over a-spatiality.

Phase I: vocal mobile communication

In 2001 the cultural geographer Eric Laurier published an article entitled *Why people say where they are during mobile phone calls*, an ethno-methodological and conversational investigation that examines a feature of mobile phone conversations–such as providing verbal geographical information or, according to Emmanuel Schegloff, "locational formulations" as part of the opening sequence of a phone call.

It is clear to Laurier that this information, besides being used to coordinate a possible face-to-face meeting (micro-coordination needs, according to Richard Ling, 2004), supplies a relevant *context* to whether we can talk about what we want on the phone or not; who we can expect to be overhearing our calls at the other end, how long the call can be, what it might be interrupted by, etc.

To summarize, the socially competent user needs to pay attention to the contextual sensitivity of the conversation, since any information on the physical location can be useful to define the social situation in which the conversation occurs, and in this way to determine the conditions of the communication itself (availability to communicate, privacy level, duration of call, themes, tone, etc).

So Laurier reminds us that in mobile telephone conversations the physical space and the social situation become one of the relevant contents of the communication, because of their capacity to condition the virtual space of flows.

Phase II: visual mobile communication

With the social diffusion of camera phones, starting more or less from 2003, we have seen the production and exchange of images at a distance, a practice that has become ever more popular starting from the recent facility to share pictures in web 2.0 (Scifo, 2009).

In 2004 I personally conducted (at the OssCom - Centre of media and communication research based at the Catholic University of Milan) an ethnographic research on the early process of domestication of camera phones and on the new mobile visual culture of sharing snapshots among young Italians (I interviewed 70 subjects aged 14-35).

We could now ask, quoting Laurier, "Why do people show where they are in mobile visual messages at distance?" Our main findings (Scifo, 2005) indicate that visual communication at distance seems to act firstly as a form of testimony and authentication of one's presence in a certain physical space, to show the group of peers. The typical "Look where I am" (as one interviewee put it: "you say: 'Look, there she is! And here I am!'"); secondly as a tool for sharing that same space and the objects that may be there:

> I went to a fair in Germany, in March, and I took pictures of what I could see, and sent them to my friends; if a friend tells me about something that he is watching, I can tell him "show it to me".

In this sense, the visual message is used for witnessing and describing.

When on the other hand it is the (individual and social) situations and emotions that one wishes to share, the intended aims can be multiple, albeit interrelated. For instance, to narrate what one is doing and experiencing in one's immediate space and time:

> The picture you take while you're dressing and diving, and then you send it by MMS, it's fantastic! It's fun! … My friends are performing in a club, they take a picture and send it to me.

Or one might be trying to include absent friends in a face-to-face social interaction. This is particularly evident in the following statement by a girl:

> When she's working at the hospital, when she's on duty in the afternoon, in the evening, or at night, and I go out with my friends and she's not there with us because she's working, I can take a picture of my friends and send it to her, so she can join us in some way.

Often this inclusive aim is coupled by the desire to make fun of each other, teasing those who are not sharing the same space and time:

> "Hi! Eat your heart out, we are skiing!" - a way of saying "I am here having fun, and you are not".

So the remote, simultaneous visual communication among young people is becoming a way for them to take hold of the experiences and worlds of others:

> For instance, the emotion of having a child... he is the father, but I lived it too; the opportunity to see places I could not see, because I was elsewhere…

Visual messages are seen and interpreted as a way to extend and multiply the experience.

Finally I will briefly talk about another useful finding related to the next topic. We have found a use of MMS not only at a representative level as seen up to now, but also as a true linguistic resource, effectively replacing text messages or telephone calls. In these cases the visual messages are interpreted as a possibility to enable action, social co-ordination, and problem management:

> Sending a picture showing yourself having lunch at a certain bar, so that if someone wants to join you, that is the place he should head to.

An instrumental use of mobile visual communication (equalling text messages and vocal calls) to organise circumstances of physical proximity, associated with just-in-time, flexible last-minute coordination, also on the move.

All of the above findings show how the experience of visual communication is closely linked to the search for a spatio-temporal embedding of subjects and experience. The context of communication, physical location and social situation, as the exclusive content of the virtual co-presence space, becomes text.

Phase III: geo-localized communication via mobile internet

Since 2008 geo-localization services, which were made possible by the convergence of Internet, mobile phones and GPS technologies, have become a commodity through the diffusion of smartphones first and tablets later. In this paper we only consider the experience made possible by social networking applications, such as Facebook Places, that, stimulated by the success of Foursquare, have incorporated localization functions. They enable users to “check-in”, that is digitally confirm their physical presence at particular place. These functions have the purpose to show other members of our network where we are and with whom.

Reformulating again Laurier’s initial statement, the last question we ask ourselves is: “Why do people localize themselves when they access a social network from a mobile device?”.

We try to give some provisional answers, starting from the early results of an explorative ethnographic study on adolescents’ use of Facebook via mobile internet (the data were collected in June from 24 teenagers), that I am conducting with other OssCom colleagues (Mascheroni, Murru & Scifo, 2011). Both individual and group interviews have been made, in order to explore both social representations and practices related to social location-based applications.

At a phenomenological level the collected data supply us with a series of social meanings connected to the practice of localising oneself that–as we will see–come very close to the meanings attributed to the different forms of mobile communication, especially the visual communication at a distance (the findings related to the social meanings are valid also when the locative service is personally not used or judged to be “useless”, “ridiculous”, “for people who have nothing else to do”, “I’m not interested”, “who cares where they are”).

First of all, the subjects’ location is made visible to their network, using it as a symbolic resource for identity, to be socially displayed and shared:

> Because they want to show that they’ve been to that particular place; it’s just an image thing; to show they were together.

A practice that at the moment seems more connected to the out-of-the-ordinary experience of being in that determined place, but that certainly meets a need for embedding and contextualization of identity:

> Sometimes I use it, it depends on the place where I am, if I’m by the sea or at Gardaland [a famous Italian theme park], these special places; then it depends on where you are, not if you’re just in a cafe! If you’re at a Formula One race, yes.

Secondly, they find the need to communicate the location because implicitly this information indicates the social situation of the subject and establishes the limits of his/her communicative accessibility ("I'll check in at University so they know they can't disturb me...").

Next, socialising one's location on Facebook can also make it possible to coordinate the action and interaction, replacing a text message or a phone call, as stated by a boy:

> I was there waiting for him outside University and I didn't see his scooter, I didn't know if he'd already gone away, if I'd had the iPhone I'd have used Facebook places.

Moreover, publishing an updated location status may foster unplanned face-to-face meetings. So it is a practice oriented towards increasing in-presence social interaction with people from the user's social network. This being a completely new practice with respect to traditional mobile communication. The following is how a girl imagines its use:

> Like I'm here, and there's someone who just got off the closest subway stop, at 7 I'm leaving, and maybe he's getting back to the subway, he sees I'm here and says: "oh do you want to walk together?"; so if someone's interested in you, he'll see it and go there.

Finally, to "check in" to one's own Facebook page also means to activate a virtual form of appropriation of a new place:

> I use it, for example I've been to New York and then I checked in.

From this early data, we can see that with the geo-localization services associated to social networks the cycle is completed, so that the physical context, the places we pass by daily and/or "live in" automatically become embedded in the text of virtual communication flows.

Conclusion

In contrast with the sociological debate that emphasizes the processes of dis-embedding and de-localization of the subjects' experience from the social context due to the increase of both physical and virtual mobility, and the consequent weakening of a "sense of place", the empirical evidence illustrated so far shows that the mobile communication technology actually activates forms of experience that are strongly embedded in the physical space. In this way the subjects' need to be rooted in localized and socially contextualized forms of interaction is shown.

In this regard, following the reflections made by Paddy Scannel regarding radio and television and recently picked up by Shaun Moores (2003) in a broader context, we can affirm that mobile media also seem to enable "the doubling of place" (which is translated in the experience of being in two places at the same time, in the physical *here* and the communicative and mediatic *there*) or, more accurately, the *pluralisation* of places (and experiences) and their *trans-location,* through the multiplication of connections between different physical and social spaces: quoting Couldry e McCarthy "media, as we use them, multiply the possible situational connections between places […] the emerging picture, then, is not the collapse of place […] but instead the more subtle integration of our interactions with other spaces and agents into the flow of our everyday practice and experience" (2003, p. 8).

From this point of view localized social networks do not represent a novelty as such, that is a "newly found" link between new media and real places, but they simply *highlight* this link. More accurately, a communication that is ever more localized and localising is certainly increasing the *location awareness* of the subjects, as claimed by Gordon and De Souza e Silva (2011) in their recent "Netlocality" book.

Moreover, as seen in the last empirical findings, the socialization of local setting in the digital space can become a tool that modifies and/or increases the forms of interaction in presence of the social actors that use location based communication in the public space.

Finally, an interpretative hypothesis, needing further empirical verification, is related to the possible re-definition of the relation between forms of physical and virtual mobility and the so-called "sense of home". It has been observed (Silverstone, 1999; Moores, 2007) how also the use of new media has the potential to become known, 'concrete' places (Tuan, 1977) in the course of frequent use, and may contribute to creating a sense of "at-homeness". As a matter of fact, despite the increase of mobility there is still a need to embed personal and intimate relationships and our "at-homeness" in *places*, albeit virtual and de-materialized (for example in our own phone book, SNS profile, blog pages, etc.). This contributes to what has been called the "dislocation of domesticity" by David Morley (2003). New symbolical and social places that foster re-territorialisation processes of social relations in the virtual space, in any case a point of stability and return. However, today the tension that is shown through localization practices in the digital environment seems to underline a contrary, or at least complementary need: to "mobilize" also these spaces, our virtual homes, and make them ever more connected with the off-line world and embedding them in the plurality of real places that we pass by or live in our daily lives, thus fostering new links between co-presence, mobility and communication.

References

Castells, M. (1996). *The Rise of the Network Society*. Oxford: Blackwell.

Couldry, N., & McCarthy, A. (Eds.). (2003). *Media/Space: Place, Scale and Culture in a Media Age*. London: Routledge.

Gordon, E., & De Souza e Silva, A. (2011). *Net-locality. Why locality matters in a networked world.* Oxford: Wiley-Blackwell.

Harvey, D. (1989). *The Condition of Postmodernity*. London: Basil Blackwell.

Laurier, E. (2001). Why people say where they are during mobile phone calls. *Environment and Planning D: Society and Space*, *19* (4), 485-504.

Ling, R. (2004). *The Mobile Connection: the cell phone's impact on society*. San Francisco, CA: Morgan Kaufmann.

Mascheroni, G., Murru, M., & Scifo, B. (2011, September). *Mobile social networking among Italian teens: preliminary findings*. Paper presented at the final Conference Eu Kids Online, London.

Meyrowitz, J. (1985). *No Sense of Place: The Impact of Electronic Media on Social Behavior.* New York: Oxford University Press.

Moores, S. (2003). Media, Flows and Places. *Media@lse Electronic Working Paper Series*, No. 6, Department of Media and Communications, London School of Economics and Political Science, London.

Moores, S. (2007). Media and Senses of Place: On Situational and Phenomenological Geographies. *Media@lse Electronic Working Paper Series*, No. 12, Department of Media and Communications, London School of Economics and Political Science, London.

Morley, D. (2003). What's 'Home' Got to Do with It?: Contradictory Dynamics in the Domestication of Technology and the Dislocation of Domesticity. *European Journal of Cultural Studies*, *6* (4), p. 435-458.

Scifo, B. (2005). The Domestication of Camera-Phone and MMS Communication. In K. Nyìri (Ed.), *A sense of Place. The Global and the Local in Mobile Communication* (pp. 363-373). Vienna: Passagen Verlag.

—. (2009). The Sociocultural Forms of Mobile Personal Photographs in a Cross-Media Ecology: Reflections Starting from the Young Italian Experience. *Knowledge, Technology & Policy*, *22* (3), 185-194.

Silverstone, R., Hirsch, E., & Morley, D. (1992). Information and communication technologies and the moral economy of the household. In R. Silverstone & E. Hirsch (Eds.). *Consuming Technologies: Media and Information in Domestic Space* (pp. 15-31). London: Routledge.

Silverstone, R. (1999). *Why study the media*. London: Sage..

Tuan, Y. F. (1977). *Space and Place: The Perspective of Experience*. Minneapolis, MN: University of Minnesota Press.

CHAPTER VIII

THE MUTED MOBILE IN TOKYO

SATOMI SUGIYAMA

1. Rationale

The way people interact with each other is fundamental in considering the notion of cities. According to Mumford (1937 [2000], p. 29), the city is "a related collection of primary groups and purposive associations" and "a geographic plexus, an economic organisation, and institutional process, a theatre of social action, and an aesthetic symbol of collective unity". The city, then, can be conceived of as a complex theatrical stage on which people with varied relationships and social roles engage in everyday interactions. In this sense, examining the norms of social interactions is a critical endeavour for understanding the city.

In this endeavour, a consideration of media is pivotal because mediated social interactions have become an essential aspect of the symbolic interactions that people engage in the city. Rheingold (2002) succinctly pointed out the impact mobile media has been exerting on the city. He states, "(s)warming supported by texting and mobile telephony, untethered ubiquitous Internet access, location-aware services, and device-readable information associated with specific places are only the beginnings of significant changes in the way people use urban spaces" (Rheingold, 2002, p. 206).

Earlier studies of the mobile phone (e.g., Fortunati, 2002; Katz & Aakhus, 2002; Ling, 2004) noted how the mobile phone brought a new kind of disorder into public spaces. From ring tones to telephone conversations in a variety of social arenas, past research reported how people are disturbed by the "noise" the mobile phone introduced, leading to the discussion of appropriate and inappropriate mobile phone use in public spaces.

In Japan, mobile phone use was already considered a disturbance of public space in the early 1990s (Matsuda, 2005, p. 23-24). At the beginning of the 2000s, *chakumero* (ring tones) and *chakuuta* (ring songs) flourished (see Okada, 2005, for more discussion), potentially turning the "noise" into "a pleasant experience and even a treat to bystanders" (Licoppe, 2005, p. 149). In this way, the mobile phone's ring can be reshaped as "an act of 'positive politeness'" of the Brown and Levinson's politeness theory, and can be also considered as a form of self-expression (Licoppe, p. 150). The popularity of these ring tones and songs, however, did not last. What is noticeable observing major Japanese cities such as Tokyo nowadays is that the sound-based "noise" has been largely erased in the public spaces of Tokyo. Tokyo is a very populated city (more precisely, a metropolis) with more than 13 million people (Statistics of Tokyo, December 2011). As Rheingold (2002) describes, about 1,500 people cross the multi-directional Shibuya intersection every time the lights turn green. The Shibuya crossing is said to have "the highest density of mobile phone use in the world" (Ito, 2005, p. 141). Yet, this media rich and highly populated city is remarkably quiet for a city where so many people with varied relationships, from complete strangers and bystanders, familiar strangers (Milgram, 1972 [1992]), to intimate others, interact.

It is within this context that the mobile media device is used and emerges as a symbolic object. As the mobile phone continues to develop into a device that seems no longer appropriate to call a telephone, norms of behaviours involving the mobile device have also been evolving. This paper explicates how people are experiencing this ever-pervasive mobile device in the city of Tokyo. The purpose of the study is not to discover the novel behaviours involving the mobile device, but instead, to offer some data to consider how mobile behaviours have changed or have not changed over the decades. Through examining a case of college students and young professionals, the present paper seeks to conceptualize the current trend of how people use the mobile device and how such norms are influencing the city of Tokyo.

2. Method

In order to examine the proposed question, the data from focus group interviews conducted in Japan during the summer of 2010 were analysed. The method of focus group interviews has its strength in creating a dynamics that is similar to one of everyday social discourse (Lindlof, 1995). Four sessions of focus group interviews were conducted. In order to keep each focus group demographically homogenous (Lindlof, 1995), each group was composed of participants with similar backgrounds in terms of their age and professions. More specifically, one group was composed of young professionals, between the ages of 35 and 40, who work at the major corporations in Tokyo (FGI 1), and two groups were composed of undergraduate students from a private university in Tokyo (FGI 2 and 3). The study also draws some data from a focus group that was composed of graduate students in the Master's program at a public university outside the Nagoya area (FGI 4). Since they do not reside in Tokyo, the data was used as far as it complemented the first three focus groups. In total, 17 people participated to the focus group interviews. In addition to the focus groups, two junior high school students and one business entrepreneur were interviewed as a supplemental data source. However, this portion of data is not reported in the present paper.

Each interview session lasted about one hour, taking a semi-structured approach. The interview schedule was adopted from the previous study conducted in 2006 (Sugiyama, 2006). The interview began with some questions regarding when they started using a mobile phone as well as how often and in what ways they use it. It then moved on to some questions about the style and the decoration of their mobile and what the mobile means to them. All focus group interviews were audio-recorded and transcribed.

Once the transcriptions were completed, the data was analysed using the grounded theory approach (Strauss & Corbin, 1998). In this process, the past literature and experience were used to sensitise the analytical perspective of the researcher, while minimising the possibility of forced interpretations. Throughout the analysis, the researcher sought to discern phenomena, which is "repeated patterns of happenings, events, or actions/interactions that represent what people do or say, along or together, in response to the problems and situations in which they find themselves" (Strauss & Corbin, p. 130). In the present study, the researcher focused on some repeated patterns of the norms of their mobile use in the city in order to understand the aforementioned question.

3. Findings

3.1 Talking on the phone

It is well documented that texting has been widely adopted as a means of mobile interaction rather than telephone conversations, particularly among young people (e.g., Baron, 2010; Matsuda, 2005). The present case of focus group participants was not an exception. For instance, the focus group of young professionals said:

> F1[1]: Somehow, because the *meiru* (*keitai* email)[2] is so convenient that with those I used to chat on the phone, I no longer chat with them on the phone.
> Others: Right!
> F1: If the person is very close, I would talk (on the phone), but with regular friends…I think that with you (for F2), we used to chat on the phone saying "let's go shopping!"
> F2: Right, we don't chat on the phone any more.
> F1: Yeah, if we say "let's chat on the phone" now, I might not be able to do so! (laugh)

F1 and F2 are friends from college, and they were reflecting on how their mode of mediated interaction has changed since then. F2 further commented:

> There are so many people with whom I interact with *meiru* but not with phone calls. (I fear) calling might be bad, inconvenient for them. Receiving a call is sometimes troublesome as well. When I am relaxing, it is too much trouble to answer a call. Then, I receive *meiru*.

This comment suggests her keen awareness toward the social distance with each relational partner. Whether they feel comfortable calling up and talking on the phone is a great indication of their intimacy. Furthermore, being considerate to others is often regarded as a sign of maturity in Japan. Perhaps this is a personal quality that is expected in many other societies, but what is regarded as a considerate behaviour in Japan involves a lot of attention and care for the unspoken behaviours of others. As young professionals, they are highly aware of this social expectation, and this

---

[1] F indicates female participants and M indicates male participants in the focus group. When more than one female and male participant appear in the given quote, a numbering system such as F1 and F2 indicates the range of ideas expressed by different participants.

[2] For explanations for *meiru* (*keitai* email) and SMS, please see Matsuda (2005), p. 35.

awareness potentially widens social distance that they tend to maintain with numerous others they interact with at various relational levels.

The college students from focus group 3 also made a point that they rarely use their mobile device to make telephone calls. Other than one participant who said that she talks on the phone when walking on a dark street, they reported that they "talk" on the phone only for a very short duration (e.g., 2-3 minutes to figure out the meeting logistics) or some special occasions (e.g., receiving a call from a old friend who lives at a distance).

This trend of not engaging in the audio-based phone conversation apparently contributes to the aforementioned quiet ambiance of Tokyo.

3.2 Texting behaviours

In line with past research (e.g., Baron, 2010), texting was quite common among all the focus group participants. Social interactions via mobile texting involve both interaction with co-present others/bystanders and interaction with their relational partners on the other side of the line. Across focus groups, participants reported how they engage in the texting behaviour at random moments in their everyday life in the public arena such as on the train. Participants also reported on how they witness numerous others texting, for instance, on the morning commuting trains. This suggests that mobile devices are visually present in the city.

Not only to co-present others, texting behaviours seem to be visually quite significant in interacting with others on the other side of the line, particularly for young women. Young female professionals in focus group 1 discussed how subtle use of emoticons and colours constitutes "*kawaii* (cute) *meiru.*" One of the participants commented how she gets impressed when she receives such a *kawaii meiru*.

College students in focus group 3 had an extended conversation on the appearance of *meiru.* First, they discussed how *meiru* should have proper line spaces so that it does not look too dense and "totally black." One of the female participants commented that her *meiru* to her mother tends to look totally black without emoticons. She further commented that sentences tend to end with a period or exclamation mark when she texts to her mother. On the other hand, when she sends *meiru* to her friends, she said that she never fails to use emoticons. In response to the question asking what kind of emoticons they use, the following conversation took place:

M: I use….like a *neko* (cat)[3].
All: Yeah, a *neko*! *Neko* is convenient!
F1: *Hiyoko* (chick) is also convenient!
(Others get excited showing agreement.)
Moderator: Why are a *neko* and a *hiyoko* convenient?
All: I wonder why….
F1/M: After a question mark, put a *neko*.
Moderator: A *neko* after a question mark?
F2: Becomes a bit cuter…*maroyaka* (milder).
M: Yeah, *maroyaka*! It creates a feel of *odayaka* (calmness).
Moderator: I see, like *odayaka*.
M: When I want to express *funwari* (softness)…
Moderator: *Neko* and *hiyoko*?
F1: Yes.
F2: We can use them for any occasions.
F1: When I'm happy or I'm a bit troubled, it's like "well, let's use them anyhow at the moment"
M: For anything (laugh)
Moderator: I never thought a *neko* is convenient.
(All laughed)
F2: Also, a *kirakira* (sparkle) is convenient.
F: Yeah, a *kirakira*!

What the conversation suggests is that these participants are keenly aware of the impressions that the *meiru* visually conveys to their interactants. They all knew that both a cat and a chick are yellow, and are displayed across the mobile device of different manufacturers, unlike a rabbit, a frog, or a turtle. Although this keen awareness seemed to be more prominent among young women, the college students reported their perception that men are using more emoticons these days. In fact, the male student in the quote above was fully participating in the conversation, agreeing with female students. They commented that men are becoming softer and rather "herbivore" citing the recent popular media characterization of Japanese men.

Although the appearance of the mobile texts is not normally visible in everyday interactions in the city, it offers interesting insights into the texting behaviour people engage in around the city. The extent to which

[3] The original Japanese words are kept here because their English equivalents lose a lot of nuances conveyed in Japanese.

these participants are concerned about the image they convey to their relational partners draws their focus to the mobile device and their relational partners rather than the co-present others and bystanders in the city. They are engaging in the "telecocooning"[4] behaviour (Habuchi, 2005). What the present data highlight is how neatly and carefully they manage to maintain the pleasant mood and avoid relational disturbance with their relational partners on the other side of the line as well as with co-present others and bystanders as they engage in texting behaviour in the city.

3.3 Web-browsing behaviours

Another type of mobile behaviour that the focus group participants frequently commented on is web-browsing. For example, in all focus groups, they reported that they often browse the *mixi* site, a Japanese version of a social networking site. A male student in focus group 2 said that he has joined various communities for his personal interests such as music, comic books, and sports, and browses those sites frequently. Female students in focus group 3 reported that they follow their friends via twitter and *mixi*, as well as browsing the blogs about restaurants. Just like texting behaviour, this way of using the mobile device makes the device quite visible while keeping it quiet in the city.

3.4 The mobile appearance

The appearance of the mobile device, including its style and colour as well as its decorations such as charms and stickers, has been documented in the case of Japan (e.g., Sugiyama, 2006, 2009). This is a form of a social interaction involving the mobile device since people perceive and express the self and social connections (Sugiyama, 2009). This aspect still seems to be quite prevalent. The young professional in FGI 1, who commented about *kawaii meiru*, had a cute strawberry charm on her mobile device because she likes strawberries. College students in the focus groups reported that they attach charms that match those sported by their friends and club members.

---

[4] According to Habuchi, telecocooning is "a zone of intimacy in which people can continuously maintain their relationships with others who they have already encountered without being restricted by geography and time" (p. 167).

The mobile decoration was also reported as a significant public display of intimate romantic relationships. Participants of focus group 3 commented on how women attach the same decorations as their friends, while men do not. When a man has a matching decoration, it is usually matching with his girlfriend's. In particular, they reported that it is a trend for a boy to attach a charm of Mickey Mouse and for a girl to attach a charm of Minnie Mouse, or a version of such (e.g., Donald Duck and Daisy). This can also visually suggest that the couple went to Disneyland together. For these college students, the couple does not have to be there together holding hands and kissing[5]. They don't have to verbally state that they went to Disneyland or they have a boyfriend/girlfriend. It is obvious just from the appearance of their mobile device.

Just like the intimate connection can be quietly conveyed via the appearance of a mobile device, when the intimacy gets into trouble as well, the troubled relationship becomes quietly exposed:

> Moderator: So, if a boy has a matching charm, it is matching with his girlfriend?
> M: Right.
> F1: And if it is removed…
> F2: Yeah, just from looking, it's like… I see…
> M: It's like… ah…
> Moderator: It becomes obvious.
> F3: Yeah, that kind of charm… I always keep it attached.
> Moderator: So if it's gone…
> F2: It's like… they broke up?
> Moderator: Like others feel somewhat uncomfortable?
> F2: Yeah, like… I don't know what to do…
> F1: Yeah, like s/he had that many matching ones!
> Moderator: Does such a thing actually happen?
> F2: Yeah.
> F1: Quite often.
> F4: It's so obvious.

As the conversation shows, the appearance of the mobile device conveys relational status quite vocally, suggesting that their relational status does not have to be "voiced" out loud, but instead, can be eloquently conveyed visually in public.

---

[5] Public displays of romantic affections are not very common in Japan.

3.5 Ringtones

Interestingly, all focus group participants of the present study reported that they keep their mobile device on silent mode when they are not at home. Some of them keep their mobile device on silent mode even at home. In all focus groups, the participants reported that a primary reason for keeping their mobile device on silent mode is to avoid an accidental ring in class, at work, or in public places such as on the train. Those who keep their mobile device on silent mode even at home mentioned that it is troublesome to switch back and forth between the sound-on mode and sound-off mode. For example, the following conversation took place during the focus group interview 3:

Moderator: Do you use a ring tone?
All: No.
M: I use the manner mode.
F1: I use the silent manner.[6]
F2: All the time? Even at home?
F1: Yeah, I use the silent manner even at home.
M: Me too, just vibe.
F3: Do you notice when you receive (a *meiru* or a call)?
F1: No.
Others: That's what I thought! (laugh)
F2: That's why you have to carry it in your hand.
F1: Ah, maybe! It just flashes.
M: You can never answer a phone call.

According to them, most college students keep their mobile on silent mode, and they cannot remember anyone who uses any songs or particular melodies. This suggests, in the case of Japan, that the mobile phone's ring is not considered as a treat for the bystanders, but as an annoyance and disturbance regardless of the music selection. They actually commented that they are bothered when they hear someone's *chakumero*. One stated, "Wherever I go, when *chakumero* rings, it's like annoying. When I hear other people's *chakumero*, I think it's noisy," and others agreed.

Adopting Licoppe's aforementioned discussion on ring tones, this can be interpreted as a rejection of self-expression in a public space. Because

[6] Both manner mode and silent manner refer to the silent mode. These terms are used in Japanese as they are (pronounced in the Japanese way). It is certainly interesting to note how the notion of "manner" has been closely associated with the silent mode.

the participants are very aware of the prevailing sentiment, they keep their mobile on silent mode, or rather *muted*, so that it cannot pronounce its presence to the public. If we take the metaphor of the mobile as an extension of the body and representative of the self (e.g., Fortunati, 2003; Fortunati, Katz, & Riccini, 2003; Fortunati & Vincent, 2009; Katz, 2003; Oksman & Rautiainen, 2003a; 2003b; Sugiyama, 2006, 2009, 2011), the *muted* mobile can also suggest that these participants are muting themselves so that they will not pronounce their presence to the numerous co-present others in the city.

3.6 Placement of the phone

It appears that those who always keep their mobile device on silent mode tend to keep their mobile at a visible place so that they can constantly attend to their mobile and will not miss incoming contacts. For example, the female student in the previous example reported that she usually keeps her mobile in her hand. In fact, all female participants of focus group 3 agreed that they walk in public with their mobile in hand.

Another example can be found in the following statement by a focus group 4 participant:

> M: In principle, I keep my mobile on a place I can see, such as on the table. In this way, I can see it flash when I receive something.
> Moderator: So rather than putting it in your pocket…
> M: Place it on a visible spot.

All the participants to focus group 4 reported that they carry their mobile in their pocket. This seemed to be a trend for all male students across focus groups. However, the one who always keeps his mobile on silent mode made the comment above. Furthermore, he stated:

> M: There was a time that I forgot to switch my phone to the manner mode, so in order to avoid my phone ring in class, I keep it in manner mode all the time. […] I don't think that I will receive any emergency phone call, and also, I think that once every 10 minutes, or at latest, I will notice within an hour. So I feel it's ok. […]
> Moderator: Do you check your phone for new messages all the time?
> M: (thinks briefly) Yeah, I do.

Perhaps because their mobile device does not ring or even make a rattling sound, they feel the need to actively check their mobile. This could be conceived of as a way of controlling not only their relationships but also their mobile. Their mobile device often brings in relational troubles at random moments, and some feel as if it is an animated object blurring the line between the mobile device and their relational partners (Sugiyama, 2011). For this reason, their animated troublesome mobile device needs to be *muted* so that they can manage their relationship with this rather rambunctious device.

4. Conclusion: the *muted* mobile in Tokyo

The present study suggests that people are very much engaged with their mobile device, but all the sounds coming from the mobile device seem to be largely disappearing in the city of Tokyo. The data presented here can be summarized in Table 1 below[7]:

Table 1 - Summary of the data

Common mobile behaviours	Physical presence of interactant(s)	Relationships with interactant(s)	Sensory mode	Trend
Telephone conversation	Co-present	Strangers/bystanders	Audio	-
Ringtones	Co-present	Strangers/bystanders	Audio	-
Web-browsing	Co-present	Strangers/bystanders	Visual	+
Texting	Co-present Absent present	Strangers/bystanders Relational partners	Visual Visual	+ +
Mobile appearance	Co-present	Strangers/bystanders Relational partners	Visual Visual	+ +

What the table illustrates is the trend that the behaviours that involve the mobile device are becoming less audio-based, and instead, becoming more visual-based in Tokyo. As numerous scholars (e.g., Hofstede, 1980, 1991; Triandis, 1995) described, Japanese culture highly values social harmony, and social disturbance is heavily sanctioned. Although these cultural characteristics that developed decades ago need to be approached with caution, the way the focus group participants explained their mobile behaviours suggests that these cultural traits are still quite ingrained in the norms of everyday social interactions in Japan. Baron's cross cultural studies of Sweden, U.S., and Japan (2010) also suggest that the people in Japan are under a great pressure not to disturb others in public. The mobile telephone that started to pervade Tokyo in the Nineties was regarded as a

[7] This is not meant to be a comprehensive list of all possible interaction patterns. It merely summarizes the points discussed in the previous section.

noisy medium and social rules started appearing in numerous public places including trains (Matsuda, 2005). The culture of "being considerate" seems to have facilitated self-monitoring so that people do not have to feel socially sanctioned in public. Referring to the aforementioned Shibuya intersection, Rheingold (2002) states, "1,500 people cross from eight directions at once, performing a complex, collective, ad hoc choreography that accomplishes the opposite of flocking; people cooperate with immediate neighbours in order to go in different directions" (p. 2). Perhaps this "complex, collective, ad hoc choreography" performed on the multi-faced theatrical stage of the city is a backbone of the currently observable behaviour that involves the mobile device.

This quiet and "considerate" use of the mobile device could also suggest how people are *forced* to take part in the choreography in a very socially savvy manner. The extent to which the focus group participants express their worry for their mobile phone's ring at inappropriate places hints at such a pressure. Some of them also expressed their serious concerns for disturbing others when their mobile rings, and also, their wish for not being disturbed by their own mobile rings triggered by others. This cultural pressure might suggest that the mobile device is not only *muted* in terms of its actual sound, but also in terms of the metaphorically toned-down self-expression. For this reason, the *muted* mobile could connote a way of domesticating the medium (Silverstone & Haddon, 1996) that interrupts a calmly and neatly choreographed theatrical performance, while somewhat sacrificing the opportunity for self-expression and perpetual contacts (Katz & Aakhus, 2002) taking advantage of the sound-based telephone conversation, richer in communicative cues than text-based interactions. At the same time, it is notable how the mobile device is visually quite present in urban public places. In addition, the attention exerted to the visual appearance of the mobile device and the *meiru* signifies the importance of the visuals involved in mobile behaviours. The *visually vocal* norms of behaviour surrounding the mobile device could be complementing the *muted* mobile device. The dynamic interplay between the *muted* aspect and the *visually vocal* aspect might capture the culturally appropriated norms of the mobile behaviours in Tokyo at the moment.

References

Baron, N. S., & af Segerstad, Y. H. (2010). Cross-cultural patterns in mobile phone use: Public space and reachability in Sweden, the USA, and Japan. *New Media & Society*, *12* (1), 13-34.

Fortunati, L. (2002). The mobile phone: Toward new categories and social Relations. *Information, Communication & Society*, *5*, 513-528.

—. (2003). *The human body: Natural and artificial technology*. In J. E. Katz (Ed.), *Machines That Become Us: The Social Context of Personal Communication Technology* (pp. 71-87). New Brunswick, NJ: Transaction.

Fortunati, L., Katz, J. E., & Riccini, R. (2003). *Mediating the Human Body: Technology, Communication, and Fashion*. Mahwah, NJ: Lawrence Erlbaum.

Fortunati, L., & Vincent, J. (2009). *Introduction*. In J. Vincent, & L. Fortunati, (Eds.), *Electronic Emotion: The Mediation of Emotion via Information and Communication Technologies* (pp. 1-31). Oxford: Peter Lang.

Habuchi, I. (2005). *Accelerating reflexivity*. In M. Ito, D. Okabe, & M. Matsuda, (Eds.), *Personal, Portable, Pedestrian: Mobile Phones in Japanese Life* (pp. 165-182). Cambridge, MA: MIT Press.

Hofstede, G. (1980). *Culture's Consequences: International Differences in Work-related Values*. Beverly Hills, CA: Sage.

—. (1991). *Cultures and Organizations: Software of the Mind*. Berkshire, England: McGraw-Hill.

Ito, M. (2005). *Mobile phones, Japanese youth, and the re-placement of social contact*. In R. Ling, & P. E. Pedersen, (Eds.), *Mobile Communications: Re-negotiation of the Social Sphere* (pp. 131-148). London: Springer.

Katz, J. E., & Aakhus, M. (2002). *Perpetual Contact: Mobile Communication, Private Talk, Public Performance*. Cambridge: Cambridge University Press.

Katz, J. E. (2003). *Machines That Become Us: The Social Context of Personal Communication Technology*. New Brunswick, NJ: Transaction.

Licoppe, C. (2005). *The Mobile Phone's Ring*. In J. E. Katz (Ed.), *Handbook of Mobile Communication Studies* (pp. 139-152). Cambridge, MA: MIT Press.

Ling, R. (2004). *The Mobile Connection: The Cell Phone's Impact on Society*. San Francisco, CA: Morgan Kaufmann.

Lindlof, T. R. (1995). *Qualitative Communication Research Methods,* Thousand Oaks, CA: Sage.

Matsuda, M. (2005). *Discourses of Keitai in Japan.* In M. Ito, D. Okabe, & M. Matsuda, (Eds.), *Personal, Portable, Pedestrian: Mobile Phones in Japanese Life* (pp. 19-39). Cambridge, MA: MIT Press.

—. (2005), *Mobile communication and selective sociality.* In M. Ito, D. Okabe, & M. Matsuda, (Eds.), *Personal, Portable, Pedestrian: Mobile Phones in Japanese Life* (pp. 123-142). Cambridge, MA: MIT Press.

Milgram, S. (1972/1992). *The familiar stranger: an aspect of urban anonymity.* In J. Sabini, & M. Silver (Eds.), *The individual in a social world: essays and experiments.* New York: McGraw-Hill.

Mumford, L. (1937/2000). *What is a city?* (Excerpt from Architectural Record). In M. Miles, T. Hall, & I. Borden (Eds.), *The City Cultures Reader (2nd ed.)* (pp. 28-32). London: Routledge.

Okada, T. (2005). *Youth Culture and the Shaping of Japanese Mobile Media: Personalization and the Keitai Internet as Multimedia.* In M. Ito, D. Okabe, & M. Matsuda, (Eds.), *Personal, Portable, Pedestrian: Mobile Phones in Japanese Life* (pp. 41-60). Cambridge, MA: MIT Press.

Oksman, V., & Rautiainen, P. (2003a). *"Perhaps it is a body part": How the mobile phone became an organic part of the everyday lives of Finnish children and teenagers.* In J. E. Katz (Ed.), *Machines That Become Us: The Social Context of Personal Communication Technology* (pp. 201-217). New Brunswick, NJ: Transaction.

Oksman, V. & Rautiainen, P. (2003b). *Extension of the hand: Children's and teenagers' relationship with the mobile phone in Finland.* In L. Fortunati, J. E. Katz, & R. Riccini (Eds.), *Mediating the Human Body: Technology, Communication, and Fashion* (pp. 103-111). Mahwah, NJ: Lawrence Erlbaum.

Rheingold, H. (2002). *Smart Mobs: The Next Social Revolution.* New York: Basic Books.

Silverstone, R., & Haddon, L. (1996). *Design and the domestication of information and communication technologies: Technical change and everyday life.* In R. Mansell & R. Silverstone (Eds.), *Communication By Design* (pp. 44-74). New York: Oxford University Press.

Statistics of Tokyo: http://www.toukei.metro.tokyo.jp/jsuikei/js-index.htm (accessed on January 14, 2012)

Strauss, A., & Corbin, J. (1998). *Basics of Qualitative Research: Grounded Theory Procedures and Techniques.* Newbury Park, CA: Sage.

Sugiyama, S (2006). *Fashioning the Self: Symbolic Meanings of the Mobile Phone for Youths in Japan* (Doctoral Dissertation, Rutgers, The State University of New Jersey, 2006), AAT3240282.

—. (2009). *The decorated mobile phone and emotional attachment for Japanese youths*. In J. Vincent. & L. Fortunati (Eds.), *Electronic Emotion: The Mediation of Emotion via Information and Communication Technologies* (pp. 85-103). Oxford: Peter Lang.

—. (2011, May). *Tool, partner, or a body part?: Use patterns and metaphors for the relationship between people and the mobile communication device in a case of Japan.* Presented at Seamlessly Mobile? Mobile Communication @ Crossroads. International Communication Association Pre-conference, Boston.

Triandis, H. C. (1995). *Individualism & Collectivism.* Boulder, CO: Westview.

CHAPTER IX

SERIOUS URBAN GAMES: FROM PLAY IN THE CITY TO PLAY FOR THE CITY

GABRIELE FERRI AND PATRICK COPPOCK

1. Introduction

The academic study of playful activities has rapidly gained relevance over the past ten years and given birth to an interdisciplinary field known as Game Studies. In that context, a small but promising research area deals with interactions between game design, urban planning and socially relevant issues such as urban rehabilitation, innovation, integration, inclusion and civic engagement. We believe that digital and non-digital ludic practices in urban spaces offer countless untapped potentialities for the promotion of active, responsible forms of citizenship, awareness-raising on key sociocultural and political issues and the promotion of more participative urban design and development processes.

To map a possible direction for future research, we will discuss some features of ludic activities designed to motivate players to interact with their surrounding urban space and fellow citizens in positive ways. We will describe two ludic genres: Serious Games and Urban Games, the overlapping of which defines an emergent *Serious Urban Games* sub-genre. As a first step towards a more extensive analysis, we shall then detail two key tendencies in Serious Urban Game Design using some examples to argue for the relevance and usefulness of designing, organising and studying playful activities in metropolitan areas.

1.1 The rise of Serious Games

Interactive technologies with more or less *ludic interfaces* are increasingly pervading our lives, through industrial (work consoles, robotics), entertainment (computer games, interactive fiction), information (websites, social media, electronic documentation) and communication (mobile phones, tablets) centred uses (Fuchs, 2012; Murray, 2012). Recently, a more explicit usage of interactive media for purposes other than simple entertainment is attracting more and more attention and investment: from advertising and news games to political and social commentaries, from self-motivational exercise games to critical statements regarding social causes (Bogost, 2011; Bogost & Ferrari, 2010). While the general study of ludic activities has gained significant relevance and given birth to an interdisciplinary field: Game Studies, the specific sub-genre aiming at concrete effects regarding real-world issues has been called Serious Games or Games for Change. A few outstanding examples of the above-mentioned genres are: Inside the Haiti Earthquake (PTV, 2010), IBM CityOne (IBM, 2010), ICED–I Can End Deportation (Breakthrough, 2008), PeaceMaker (Hybrid Learning Systems, 2007), The McDonald's Videogame (molleindustria, 2006), Darfur is Dying (Ruiz/mtvU, 2006).

Even if Serious Games have not yet entered the mainstream media, they are nonetheless a relevant emergent phenomenon. The Woodrow Wilson International Centre for Scholars, Washington D.C., is one of the most important research hubs advocating use of Serious Games. In 2002, it contributed to founding the Serious Games Initiative, one of the major efforts in this field, "to help usher in a new series of policy education, exploration, and management tools utilising state of the art computer game designs, technologies, and development skills". Several projects related to serious gaming practices recently obtained substantial funding. The European Union approved a budget of € 275,000 to design and develop Citzalia, a Serious Game simulating and illustrating the workings of the European Parliament. The World Bank Institute sponsored the production of the Evoke game to promote economic development in third-world countries. Finally, UNICEF in partnership with Microsoft and gameLab promoted Ayiti: the Cost of Life, focused on childrens' rights in developing countries. The influential American game scholar Jane McGonigal–lead designer of Evoke–recently published the book "Reality is Broken" (McGonigal, 2011), addressing the relevance of play-centric design for solving concrete, real-world issues.

1.2 The rise of Pervasive and Urban Games

Pervasive and Urban Games are other new topics in Game Design and Game Studies that, in general, seem to defy Johan Huizinga's well-known notion of the Magic Circle. The Dutch cultural historian defined play as

> a free activity standing quite consciously outside "ordinary" life as being "not serious", but at the same time absorbing the players intensely and utterly. [...] It proceeds within its own proper boundaries of time and space according to fixed rules and in an orderly manner. It promotes the formation of social groupings, which tend to surround themselves with secrecy and to stress their difference from the common world by disguise or other means. (Huizinga, 1938)

A similar notion is also present in Erving Goffman's (1961) concept of a "metaphorical membrane" that selects filters and transforms the events, actions, and properties existing outside the game.

Pervasive Games do not adhere to those classical definitions but share some different traits, such as not being tied to a specific place, time and technological medium and involving significant interactions to be carried out in public spaces. Yet, they are still games–argues game scholar Markus Montola–even if their Magic Circle is extended in different ways:

> The game no longer takes place in certain times or certain places, and the participants are no longer certain [whether they are playing or not]. Pervasive games pervade, bend, and blur the traditional boundaries of game, bleeding from the domain of the game to the domain of the ordinary. (Montola, Stenros & Waern, 2009, p.12)

In this paper, we will define Urban Games as a specific subset of Pervasive Games, set in metropolitan areas, which encourage participants to move freely through public areas and to socialize by interacting with bystanders. In general, Urban Games are often designed to create a minimal level of competition between players, emphasising instead exploration, experimentation and creative use of urban spaces.

1.3 Towards Serious Urban Games?

The area of focus for this work can be defined if we consider an overlapping between:

- Pervasive Games, taking place in an expanded spatio-temporal and social situation,
- Urban Games, a subset of the former genre specifically designed for urban environments,
- Serious Games, ludic systems designed for educational, critical or satirical purposes.

It is possible to combine the above categories to map the field of Serious Urban Games: playful, organised practices, taking place in urban environments with some kind of technological/digital support, and serving social purposes–i.e. raising awareness on specific issues or empowering specific minority groups. Today, this emergent field is still regarded as a niche, but if we look back at the historical development of the Serious Games genre as a whole, it is reasonable to claim that Serious Urban Games have not yet expressed more than a fraction of their full potential.

2. Two tendencies

The Serious Urban Games genre is obviously still in its infancy, and undergoing a significant evolution that will shape its future characteristics. In this early attempt at mapping two developmental trends in this area, we are aware that–while we indicate some features that in 2012 appear to define this genre–future Serious Urban Games may well introduce additional complexities.

As a beginning, two key trends will be introduced. The first will be labelled "TIAG/TINAG ambiguity": a semiotic design strategy based on the overlapping of different contextual rules aiming at mixing fictional game elements with elements from the real world. This mechanics was not born with Serious Urban Games as such, but, as will be shown, it can enhance their effectiveness. The second tendency will be called "distributed storytelling" and deals with additional tasks assigned to players, requiring them to gather narrative materials and re-tell them in order to proceed in the game.

2.1 "This is a game" / "This is not a game" ambiguity

I believe that you can reach the point where there is no longer any difference between developing the habit of pretending to believe and developing the habit of believing. (Umberto Eco, Foucault's Pendulum, p. 386)

As Huizinga, Goffmann and several other scholars have pointed out, a player of a traditional, non-pervasive game is well aware of which kind of activity he is taking part in. This frame defines the players' preliminary acknowledgements of being in a playful situation. In this paper, this outer system of player expectations will be called the *This-Is-A-Game* (TIAG) layer.

Player cooperation with an interactive system–digital or non-digital–generates a ludic discursive universe within the TIAG layer in which gaming interactions are acknowledged as fictional (Ferri, 2009). When focalization (Genette, 1972) is shifted within this universe, users abide by TIAG interpretive rules and temporarily set aside their encyclopaedic knowledge of the world–not being surprised, for example, by the height of Super Mario's jumps.

However, many other games–Pervasive Games above all–feature a second nested system of expectations within the TIAG layer. Citing the excerpt from Umberto Eco's novel Foucault's Pendulum, this second intradiegetic frame asks players to adopt "the habit of pretending to believe that this is not a game": which we can designate as the *This-Is-Not-A-Game* (TINAG) system.

The majority of Pervasive and Urban Games relies on semiotic and procedural mechanisms that generate player engagement by creating ambiguity between the TIAG and TINAG systems. The following brief examples detail the evolution of this semiotic strategy from an early casual Pervasive Game up to other, more recent, Serious Urban Games.

2.1.1 Killer/Assassin

Markus Montola and Jaakko Stenros (Montola, Stenros & Waern, 2009) recently published a short case study of "Killer", or "Assassin", a traditional game played on many American university campuses. Before continuing in this description, let us clarify that this specific game, "Killer", is not a Serious Urban Game and that we are beginning here our analysis to show how ordinary games have evolved into a new sub-genre. The true origins of this decades-old Pervasive Game are unclear. J. W. Johnson (1981) has tracked its roots to The Seventh Victim, a short story written by Robert Sheckley in 1953, and especially the Italian cult film based on the story, *La Decima Vittima* (1965). Montola and Stenros explain that movie and the game:

> It is a science fiction story about a future society where human hunts are staged, where participants alternatively adopt the parts of hunter and prey, killing each other as a part of a competition. After the film was shown in the United States, Killer games started to emerge in university campuses around the country. The game emerged as oral folklore; countless variations still exist with various names. When Steve Jackson Games codified [those] games in Killer: The Game of Assassination in 1981, the rich oral tradition was condensed on paper, listing dozens of options on how to play the game. (Montola, Stenros & Waern, 2009, p. 3)

The general rules may be summarized as follows:

-players impersonate contract killers using toy weapons and other props;

-each player is assigned a "target", another participant that must be "killed". At the same time, every assassin is also a target for another player. Each participant knows the identity of his "victim" but not those of the other players and, specifically, of the killer targeting him;

-the act of "killing" is simulated metaphorically using an arsenal including water guns, plastic knives, vinegar (poison), and alarm clocks (time bombs)

-the goal of the game is to score as many points as possible. Every time a player kills his victim, he scores a point. "Dead" players are removed from the game.

From a design perspective, the ambiguity between TIAG and TINAG expectations in this game exemplifies a powerful semiotic mechanism that may be used for motivating players to undertake certain actions in the real world. Through its rules and procedures, the Killer game encourages different behaviours to be adopted by players within the fictional TIAG world. For instance, paranoia may be well justified within the TIAG setting: fictional characters in the game have every right to wonder who is conspiring to kill them. In the diegetic world of Killer, carefully inspecting a cup of coffee offered by a friend or checking if somebody is hiding behind a door may be smart ideas. It is at this point that the pervasive nature of Killer becomes relevant: the "make-believe" part of this game is very limited, and the rules require that every action taken inside the TIAG layer exist also in the real TINAG world. Perfectly reasonable TIAG actions–such as checking behind doors for potential assassins–must be enacted for real, creating a strange, out-of-place, second order set of meanings.

Not only does a great part of the fun in the Killer game derive from this awkwardness, but requiring players to act in real time in both the TIAG and TINAG layers is also a precious design strategy that–as we shall describe in our examples to follow–will be useful for creating Serious Urban Games.

2.1.2 Cruel 2 B Kind

Cruel 2 B Kind is a variant of Killer designed by game scholars and designers Ian Bogost and Jane McGonigal, to add a more socially-conscious theme. The designers adopted the same mechanics as Killer while radically altering the guiding metaphors for the game.

> [...] in Cruel 2 B Kind, players "kill with kindness"–each player is assigned a "weapon" and "weakness" that corresponds with a common, even ordinary pleasantry. For example, players might compliment someone's shoes, or serenade them. While Assassin is usually played in closed environments such as college dorms, Cruel 2 B Kind is played in public, on the streets of New York City or San Francisco or anywhere in the world. Players not only don't know who their target is, they also don't know who is playing! In these situations, players are forced to use guesswork or deduction to figure out whom they might target. As a result, players often "attack" the wrong groups of people, or people who are not playing at all. The reactions to such encounters are startling for all concerned; after all, exchanging anonymous pleasantries is not something commonly done on the streets of New York. (Bogost, 2007)

Cruel 2 B Kind employs a TIAG/TINAG overlapping very similar to the one used by the Killer game. However, it adopts this strategy not only to engage players in an entertaining game, but also to spark reflection on social practices. Bogost and McGonigal addressed the issue of citizens ignoring one another in public areas and designed Cruel 2 B Kind to motivate players to interact with random passersby. While the Killer game aimed at making players look paranoid, the objective for this game is to persuade participants to be more outgoing. Again, Bogost notes: "this juxtaposition of game rules and social rules draws attention to the way people do (or more properly, don't) interact with one another in everyday life" (Bogost, 2007).

2.1.3 Big Urban Game

Big Urban Game (BUG) is a Pervasive Game/Public Art project that, at first sight, may seem very different from the previous examples. However, we will show that a variant of the TIAG/TINAG semiotic strategy is also at work in this game.

BUG was organised by the Design Institute of the University of Minnesota as a part of its Twin Cities Design Celebration, with the goal of encouraging residents of Minneapolis and St. Paul to see their surroundings in a whole new way, and to think about the design of urban space. Stakenas and Zurkow (2006) summarize the event as follows: "The BUG took place over five days in Minneapolis and St. Paul. Three teams raced 25-foot tall inflatable game pieces through a series of five checkpoints, hoping to make it to the final destination in the shortest amount of time". Big Urban Game was divided into a real-world pervasive game and its online counterpart. Designer Jane McGonigal describes its two parallel worlds:

> Three thousand, three hundred and six members of the public registered to play […] online and were divided into three teams: red, yellow, and blue. Each […] online team partnered with a dozen real-world runners, who would be responsible for moving their team's […] inflatable game piece around a 108 square-mile game board. Every morning for five consecutive days, the online players studied a digital map of the Twin Cities and voted for one of two potential racing paths. Every evening, after the votes were counted, the real-world runners raced through the city streets following the route chosen by their online counterparts. (McGonigal, 2006, pp.175-176)

In the previous examples, we pointed at TIAG/TINAG strategies based on exporting TIAG practices into the real world. In this case, the same mechanism is present but works in reverse, exporting significant real-world elements into the TIAG layer. As a matter of fact, choosing the final path for the game-piece was not an easy task, as it required a good knowledge of the city or, even, an in-person tour to decide the best option. Again, Stakenas and Zurkow write:

> Because the game pieces were literally carried through city streets, greenbelts, and alleyways, players had to negotiate the pros and cons of each route, taking into account variables such as traffic jams, low lying bridges, and busy intersections. Neighborhood connected to neighborhood as the pieces traced their circuitous routes. Citizens were left wondering

what might be next in a game where game piece-as-architectural spectacle was the rule of the day. (Stakenas, Zurkow, 2006)

In other words, significant local knowledge had to be drawn from the TINAG layer and used to formulate a winning TIAG strategic decision. BUG not only featured an easily recognizable Public Art intervention (its 25-foot tall inflatable game pieces have become almost an icon for the Urban Games genre) but required players to carefully consider the features of the metropolitan areas to be traversed by the game. Though these design mechanisms, BUG attracted attention to often overlooked urban areas and persuaded players to examine them more consciously.

2.2 Distributed storytelling

The second tendency for Serious Urban Games that we want to describe in this paper deals with more narrative-centric player engagements. Without necessarily sacrificing the more game-related mechanics, some Serious Urban Games assign to participants tasks involving the gathering and retelling of local stories, anecdotes or conversations.

2.2.1 Massively Multiplayer Soba

The Tiltfactor laboratory at Dartmouth College created a large scale collaborative urban game, Massively Multiplayer Soba (MMS), premiered in New York at the "Conflux psychogeography festival" in 2008 and later replicated in other cities. The promotional materials describing the game invited participants to "grab some friends and traverse remarkable neighborhoods in New York City [...]. Talk to strangers, find clues, and fetch ingredients for a giant collective noodle party".

Players of Massively Multiplayer Soba were divided into small groups and given a closed envelope containing a map of the surrounding neighbourhood and several written clues. The primary goal was to gather the ingredients needed to prepare a dish–soba in the first run, other Asian recipes in the later ones. Moreover, since the recipes and clues were not written in English, significant multi-cultural interactions with passersby were needed. To complete the game, players had to interview foreign-speaking citizens, asking for translation help and directions. Points were awarded primarily on the basis of the complexity and depth of interactions, rather than the "scavenger hunt" style claiming of facts, objects, or territories. The design allowed for a variety of play strategies: collaboration for those who may not be competitive, or specialization for those who excel at particular kinds of challenges. Play was combined with opportunities for engagement to allow game players to mix and interact with residents in meaningful ways, while challenging preconceptions of race and language.

Lead designer Mary Flanagan explains:

> Designed to be a team-based urban game focused on culture, food and language, the game-event enticed players to the non-tourist neighborhoods of New York, combining play with opportunities to mix with new places in the city, and ultimately take a meal in a collective dinner with local people, intentionally mixing players with non-players. This design created unexpected encounters with residents while challenging concepts of culture and language through storytelling and discovery. (Flanagan, 2010, p. 50)

The MMS series of games puts game-related elements such as rules and objectives side by side with free-form interactions with other citizens and assigns equal importance to both factors. While some more formally-defined game rules provide a stable structure for the playful practice, this MMS game series is quite open to emergent, unexpected interactions between players and their surroundings.

2.2.2 Mettiti in Gioco[1]

The rules for "Mettiti in Gioco" (MIG) / "Bring Yourself into Play" were designed during workshops led by one of the authors (Ferri) at the Faculty of Communication and Economics at the University of Modena and Reggio Emilia with the objective of encouraging multicultural dialogue in smaller urban environments. A first session of the MIG game was organised[2] in Castelnovo ne' Monti (RE), Italy, in April 2011.

---

[1] The "Intercultural Urban Game" design and research workshop was supported by a grant from the Province of Reggio Emilia (Assessorato alla Sicurezza Sociale – Consulta Provinciale per l'Immigrazione, l'Asilo, l'Antidiscriminazione).

[2] The first "Mettiti in Gioco" urban game event was co-organized by "L'Ovile – Cooperativa di Solidarietà Sociale" and by the authors, with the support of UNAR (Ufficio Nazionale Antidiscriminazioni Razziali), Province of Reggio Emilia, City of Castelnovo ne' Monti (RE) and Rete Contro le Discriminazioni (Reggio Emilia).

The design process for MIG began by selecting features from the Massively Multiplayer Soba series that could be adapted to work not only in large American cities but also in smaller Italian towns. Some of the core mechanics of MIG are actually similar to its American counterparts: the game is held in an urban area, participants are divided into groups, each of which receives a sealed envelope containing the different missions to be completed. Differently from the MMS games, MIG players were also given small hand-held video cameras and asked to document their experiences during the game. The first mission to be accomplished required the translation into Italian of a recipe written in languages[3] unknown to the players, who had to ask for help from passersby who knew the languages in question. To complete the second task, players were required to collect a story set in the town in which the game was held and to retell it to a fixed video camera at the "Story Bank". While the organisers of the game had some degree of control over the first (recipe translation) mission, the selection of which local tales to collect, and who to approach to ask for these was left completely up to the players. These first two missions were designed to be complementary, requiring players to interact with both senior citizens to collect stories from the past, and with recent immigrants to translate the multilingual clues. Later on, the third and fourth missions required participants to go shopping for the ingredients mentioned in their recipes and, finally, actually cook it and share it with other participants, also those they have asked for help.

The Story Bank is an element marking a significant difference between the MIG and the MMS game series'. It is only by "depositing" their tale in the Story Bank that players were able to complete the second mission and proceed in the game. This is because it was only the Bank that could award players, in exchange for their stories, the coupons they need to go shopping for the ingredients–as part of the next challenge. The act of retelling a story to be recorded on video in exchange for a prize thus adds value to the narration itself (a good tale is precious) and this builds a more personal, direct connection between players, passersby and the urban environment in which they are playing. In this sense, participants are no longer "just players" but they also take charge of a small but significant portion of socio-cultural, place-specific story-collection, -telling and -archiving practices.

---

[3] I.e. Russian, Arabic, Urdu and also a local dialect from the Reggio Emilia region.

2.2.3 Pie' Veloci[4]

The objective of the "Pie' Veloci" / "Fast Feet" game was to promote ecological forms of transportation as part of the 2011 European Mobility Week. This Serious Urban Game was designed to combine distributed storytelling with urban exploration on public bicycles, part of a municipal Bike Sharing initiative in the city of Bologna, Italy. Its first run on September 17th 2011 was a successful experiment, actively involving a very diverse audience including numerous participants between the ages of 10 and 14. Pie' Veloci transposed the benefits of choosing bicycles over other vehicles by contextualising them in a location-based story-hunt aimed at young players. They were asked to complete an urban tour, to collect a number of stories set in specific urban areas, to interact with passersby and to reinterpret personally one of the stories they had heard.

In this Urban Game, players were given a more specific topic–bicycles and public transport–for the tales to be collected and deposited into the Story Bank. Choosing to make players explore an urban area by bicycle while asking passersby to share with them some tales regarding "green mobility" issues served two purposes: it gave participants a more structured template to follow, but it was also an attempt to engage a wider audience besides the players themselves. This was a first attempt at giving a more public dimension to the game, creating a playful activity and a meaningful visible event, also for a non-playing audience.

3. Some preliminary conclusions

Game Studies is still a relatively new thematic area for academia and interest in Urban Games and Serious Games has been growing rapidly only over the past few years. Since the overlapping of these two last mentioned areas is quite recent, and the Serious Urban Games genre is just taking its first infantile steps, our closing remarks for this paper are particularly tentative. The main objective of these notes is to present this new field as an emergent cultural space in which research and design will march hand in hand into the future, and to highlight the current state of the art with a few outstanding case studies, while pointing out plausible directions for future research.

---

[4] The "Pie' Veloci" urban game event was co-designed and co-organized together with Sottobosco.info, with the support of the City of Bologna, Italy, and ATC, Bologna.

Our principal hypothesis is that Serious Urban Games will experience significantly increasing attention from academics, designers, public administrations, businesses, enterprises, and many other larger and smaller cultural and economic entities. To support our insights and predictions, we have examined the rise of Game Studies as a recognized academic field and Serious Games, Pervasive Games and Urban Games as relevant genres already energetically embraced by designers, policy-makers and scholars. We believe that Serious Urban Games, if they follow a similar trajectory, will rapidly become a highly relevant subject for scientific analysis and a useful tool for urban designers, planners, public administrators and actors in many other public and private culture sectors.[5]

On one hand, the specific ludic genre aiming at creating concrete social, and even political, consequences in relation to urgent real-world issues has been denominated Serious Games and is rapidly gaining acceptance and relevance amongst mainstream game designers and scholars. On the other hand, we have defined Urban Games as a particular subset of Pervasive Games, set in metropolitan areas, that encourage participants to move freely within public spaces while interacting and socialising with passersby and bystanders. Combining these two definitions, we have proposed to denominate Serious Urban Games as playful practices that aim to foster wider, concrete, social-consciousness raising effects, in players, passersby and the community at large, in a very wide range of urban and suburban public spaces, thus transforming these spaces themselves into a kind of "ludic interface",

To begin the mapping of this new and innovative area of studies and the various conceptualization and design practices that characterize it, we have examined two principal trends in contemporary Serious Urban Game design. The first we have identified as the strategic use of "This is a game / This is not a game" (TIAG/TINAG) ambiguity, and we have highlighted how this may be used to motivate players to act in the real world through use of game-based rules. The other was designated "Distributed Storytelling" which we selected in order to show how narrative-based tasks may be creatively combined with other types of game-related tasks and rules.

---

[5] For an example in connection with emergent Smarter Cities projects see http://smartercitieschallenge.org.

We are well aware that this is merely the beginning of a much longer research and design trajectory. The descriptions of the recent trends proposed here as first steps towards a mapping of the field of Serious Urban Games clearly need further analysis to identify the most defining characteristics of this genre, and our examples ought also to be compared with numerous other practices in this area. Also, more practical design work will be necessary to fully comprehend the specific features of this particular subset of urban activities. In conclusion, from these very early results, this appears to be a promising area for academic study as well as a useful, innovative tool for applied game design in urban environments.

References

Aarseth, E. (1997). *Cybertext*. Baltimore: Johns Hopkins University Press.

Bogost, I. (2006). *Unit Operations*. Cambridge, MA: MIT Press.

—. (2007). Persuasive Games on Mobile Devices. In B. J. Fogg & D. Eckles (Eds.), *Mobile Persuasion: 20 Perspectives on the Future of Behavior Change*. Palo Alto: Stanford Media.

Bogost, I., & Ferrari, S. (2010). *Newsgames. Journalism at Play*. Cambridge, MA: MIT Press.

Bogost, I. (2011). *How to Do Things with Videogames*. Minneapolis: University of Minnesota Press.

Coppock, P., & Compagno, D. (2009). *Computer Games Between Text and Practice*, EC Serie Speciale, vol. 5.

Ferri, G. (2009). Between interpretive cooperation and procedurality. In P. Coppock & D. Compagno (Eds.), *Computer Games Between Text and Practice*, EC Serie Speciale, vol. 5.

Flanagan, M. (2009). *Critical Play: Radical Game Design*. Cambridge, MA: MIT Press.

—. (2010). Creating Critical Play. In R. Catlow, M. Garrett & C. Morgana (Eds.), *Artists Re:Thinking Games*. Liverpool: Liverpool University Press.

Fuchs, M. (2012). *Ludic interfaces. Driver and product of gamification*. G|A|M|E, 1(1). Available online: http://www.gamejournal.it/ludic-interfaces-driver-and-product-of-gamification/#.Ucf2NRblk7Q, accessed: June 2013.

Genette, G. (1972). *Figures III*. Paris: Seuil; en. tr. *Narrative Discourse*. Ithaca: Cornell University Press, 1980.

Goffman, E. (1961). *Encounters. Two studies in the sociology of interaction*. London: MacMillan.

Huizinga, J. (1955). *Homo Ludens: A Study of the Play-Element in Culture*. Boston: The Beacon Press.
Johnson, J. W. (1981). *A Scholar looks at Killer: Afterword*. In *Killer: The game of assassination*. Steve Jackson Games, Austin (TX).
Murray, J. (1997). *Hamlet on the Holodeck*. New York: Free Press.
—. (2011). *Inventing the Medium. Principles of Interaction Design as a Cultural Practice*. Cambridge, MA: MIT Press.
Montola, M., Stenros, J., & Waern, A. (2009). *Pervasive Games*. London: Morgan Kaufmann.
McGonigal, J. (2006). *This Might Be a Game: Ubiquitous Play and Performance at the Turn of the Twenty-First Century*, Ph.D. thesis, University of California Berkeley. Available online: http://avantgame.com/McGonigal_THIS_MIGHT_BE_A_GAME_sm.pdf, accessed: Jan 2012.
McGonigal, J. (2011). *Reality is Broken*. London: Penguin.
Ryan, M. L. (2006). *Avatars of story*. Minneapolis: University of Minnesota Press.
Stakenas, C., & Zurkow, M. (2006). *Public Art*. Available online: http://www.o-matic.com/public_art/index.html, accessed: Jan 2012.

Games

Ayiti: the Cost of Life. GameLab, UNICEF. (2006).
Big Urban Game. Lantz, F., Salen, K.. (2002-2003).
CityOne. IBM. (2010).
Citzalia. ESN, EU. (2010-2011).
Cruel 2B Kind–a Game of Benevolent Assassination. Bogost, I., McGonigal, J.. (2006).
Darfur is Dying. Ruiz, S., mtvU. (2006).
Evoke. World Bank Institute. (2010).
Inside the Haiti Earthquake. PTV. (2010).
I Can End Deportation. Breakthrough. (2008).
Killer. Steve Jackson Games. (1981).
Massively Multiplayer Soba (series). Flanagan, M., Tiltfactor Labs. (2007-2010).
Mettiti in Gioco. Ferri, G., Università di Modena e Reggio Emilia, Provincia di Reggio Emilia. (2011).
PeaceMaker. Hybrid Learning Systems. (2007).
Pie' Veloci. Ferri, G., Sottobosco.info. (2011).
The McDonald's Videogame, Pedercini, P., molleindustria (2006).

CHAPTER X

PEOPLE, PLACES, GAMES: A MODEL TO ANALYSE LOCATION-BASED MOBILE APPLICATIONS

GIOVANNI CARUSO, RICCARDO FASSONE, GABRIELE FERRI AND MAURO SALVADOR

In this chapter we will analyse a corpus of products that are first of all hard to describe[1]. To do so, we will start from the idea of the magic circle, an extremely relevant concept among game theorists, and from there, those location-based and pervasive practices that these objects generate. Each one of these notions is borrowed from different currents of game studies and media studies, and together, they will be used to better identify our case studies. After setting up this theoretical introduction, we will present three different virtual layers–the game layer, the social layer and the narrative layer–from which we will build a model to test four case studies.

SCVNGR, *Foursquare* and *Broadcastr* are mobile applications that, despite their similar appearances involving territorial exploration and user generated contents (texts, photos, audio recording), differ in the ways their users deal with them. The fourth case, *Whai Whai* is instead a rather anomalous tourist guide–first published as a booklet and converted into an app for iPhone and iPad in 2011–which motivates its reader to explore surrounding urban space by assigning quests presenting a narrative development.

---

[1] This essay was published with minor differences as "Check-in Everywhere. Places, people, narration, games", *Comunicazioni sociali on-line*, 5, 2011. Available at: www.comunicazionisocialionline.it/2011/5/4/loadPDF/ [Accessed 11 January 2012]. An Italian version was published in the proceedings of the Media Mutations 3 conference, held on May 24-25, 2011 at the University of Bologna.

1. World exploration and location tracking

What our case studies clearly have in common is their bond with the user's displacement in the real world. In the case of *Foursquare* and *SCVNGR*, the user's movements are tracked through GPS [Global Positioning System] technologies; in *Broadcastr* location tracking is optional, while *Whai Whai* does not employ a positioning system. We will refer to those objects through the broad label of "location-based applications", although some of them do not comply with the definition proposed in 2001 by Virrantaus *et al.*: "LBSs [Location-Based Services] are services accessible with mobile devices through the mobile network and utilising the ability to make use of the location of the terminals" (2002). This definition refers mostly to satellite tracking, but in this context we will use it in a broader sense, referring to products that maintain a crucial bond with the geographic position of their user. In short, with location-based we intend both the automatic localization of the device that is running the application and/or the voluntary movements of the user following the application's spatial indications.

Some of our case studies share common properties with other products based on GPS tracking, such as interactive maps developed for mobile devices or augmented reality applications (e.g. *Wikitude*[2]). Nevertheless, we decided to focus our research on location-based applications that retain a distinct ludic component or can act as platforms for playful practices. In our paper we will explore the interactions between location tracking and games analysing both the design and the use of those objects.

2. Breaking the magic circle

In arguing that our four case studies retain a ludic or playful component, we need to confront the classic models of game studies. In defining the nature of games, most scholars have resorted to space-related metaphors. Notably, Johan Huizinga referred to the space of play and games as a magic circle. In 1938, Huizinga described it as it follows:

> All play moves and has its being within a play-ground marked off beforehand either materially or ideally, deliberately or as a matter of course. Just as there is no formal difference between play and ritual, so the

[2] *Wikitude* is an augmented reality mobile application based on GPS tracking that allows users to identify points of interest in the real world using their phone's camera.

> "consecrated spot" cannot be formally distinguished from the play-ground. The arena, the card-table, the magic circle, the temple, the stage, the screen, the tennis court, the court of justice, etc, are all in form and function play-grounds, i.e. forbidden spots, isolated, hedged round, hallowed, within which special rules obtain. All are temporary worlds within the ordinary world, dedicated to the performance of an act apart (1955, p. 10).

From Huizinga's point of view, the magic circle separates the real world from *ad hoc*, non-permanent fictional worlds that are created to play. These worlds exist within a circle, a "consecrated spot"–not unlike that of a ritual–that delimits spatially and temporally the permanence of the make-believe world of a game. Following Huizinga, Goffman claims that games are capable of creating "a locally realized world of roles and events" (1961, p. 31), while for French sociologist Roger Caillois games imply an activity that is "*sépareé* circonscrites dans le limites d'espace et de temps précises et fixées à l'avance" (1991, p. 43). For most game and play theorists then, games are inscribed in a space (physical or metaphorical) that remarks their fundamental difference from the real world and from everyday life.

Firstly, location-based playful experiences do not take place inside a physical *magic circle*, nor inside an arena designed for such activities. Their very existence depends on using the real world as a potential playground, inevitably sharing it with non-playing subjects. Secondly, while institutionalized, rule-based games require the player to comply with specific rules for the entire time of the match, while location-based products offer a less restrictive experience. The ludic component of applications such as *Foursquare* is mingled with and ostensibly depends on banal routines (shopping, driving, eating in a restaurant), while the membrane that separates play and real life remains permeable. According to game designers Katie Salen and Eric Zimmerman, while games are in the most cases formal, defined, rule-based entities, the act of playing remains inevitably fuzzy:

> The boundary between the act of playing with the doll and not playing with the doll is fuzzy and permeable. Within this scenario, we can identify concrete play behaviors, such as making the doll move like a puppet. But there are just as many ambiguous behaviors, which might or not be play, such as idly kneading its head while watching TV. There may be a frame between playing and not playing, but its boundaries are indistinct (2004, p. 94).

Christopher Moore argues that mobile gaming is defined by a ludic situation even more unstable than the one originally described by Salen and Zimmerman for conventional videogames. According to Moore, mobile games are so defined not only because they can be physically transported by the user but also because of they can fit within the player's ordinary routines:

> The mobility of play is therefore not always a series of border crossings to and from the magic circle, but a contingent process, a mode of play, seized in the movements of experience, that involves a complex relationship between different changes in time, space, social attentiveness and cultural practices that are fundamentally creative (2011, p. 378).

To face these difficulties in defining what a game is, what is a gaming experience and–most of all–what are the characteristics of location-aware playful apps, we focused our attention on objects that are pervasive in nature. To begin answering these questions, we will try to single out some of their constitutive features through the following analyses.

In a research published in 2009, Markus Montola, Jaakko Stenros and Annika Waern described a corpus of playful practices having in common "one or more salient features that expand the contractual magic circle of play spatially, temporally, or socially" (p. 12). So we are talking about something capable of transcending the game limits or, more precisely, the very idea of magic circle, and that is exactly what we were looking for. Again the three game scholars:

> The contracts of pervasive games are different from the contracts of traditional, nonexpanded games. The magic circle is not an isolating barrier distinguishing the ludic from the ordinary, but a secret agreement marking some actions as separate from the ordinary world. While all human actions are real, those that happen within the contract of a game are given a special social meaning. In conclusion, we can see that there is a twofold dynamic between the playful and the ordinary that provides pervasive games a reason to exist: Both play and ordinary life can benefit from the blurring of the boundary (p. 21).

In this work, our objective is to describe how location-based applications show a definite tendency towards blurring the boundaries that separate ludic and non-ludic activities. In the next pages we will discuss how these products elicit pervasive practices that exploit the constitutive ambiguity of the idea of playing.

3. Three virtual levels: a model of analysis

In the following pages we take into consideration the various ways in which common spatial practices (taking a stroll, going to a restaurant, exploring a new city) have become the object of an intense mediatisation. We will analyse four examples of products that aim at interacting with the user's experience of space and movement in the real world in order to test a tripartite model of analysis, based on a ludic, a social and a narrative layer.

The three layers (Figure 1) considered in the analysis of these hybrid media products allowed us to map–at least partially–the territory of space related applications, placing them within a field of tension that informs both their core design and the user experience they intend to provide.

In this perspective, the game layer refers to the presence of game-like features in the objects of our research, following the definition proposed by Salen and Zimmerman: “A *game* is a system in which players engage in an artificial conflict, defined by rules, that results in a quantifiable outcome” (2004, p. 80). We will try then to identify whether an artificial conflict, a set of rules and a quantifiable outcome are clearly recognizable in these applications. Given the recent debate around the concept of gamification and its link with marketing strategies, we propose to move past the analytic practice of singling out the ludic qualities of a software, of an app, and, instead, to begin a reflection on how they are employed. The game layer is not only about fun, it is also about the use of game features in a *factitive*[3] way with the purpose of exploiting the users' behaviours.

The social layer deals instead with the power of social networking websites and social relations to bond people together. The term retains a certain structural and ontological complexity: for Christina Prell, “social networks [are] composed of a series of levels such as actors (e.g. individuals); relations connecting actors together (e.g. friendship); dyads (e.g. pairs of actors); triads (e.g. structures composed of three actors); subgroups; and entire networks” (2011, p. 3). To dampen this complexity, we posit our analysis in the field of social networking websites. Thus,

---

[3] In his semiotic theory on modal verbs, A. J. Greimas defines the factitive modality as the object's potential skill to communicate their directions of use (communicative function) generating precise action sequences done by the users (operative function) (1983).

talking about "social layer" means to consider the connections among actors (how people are related to each other–the *social graph*) and actors and interests (as the result of online social activities: what people like, share, follow–the *interest graph*) typical of a mediated virtual networked ecosystem. Our research is not solely focused on the individuation of these tools within the analysed application, but most importantly it aims at assessing their relevance in realising a location-based ludic experience. This last consideration is especially salient in the case of *Foursquare*, a product conceived as a location-based social networking website rather than as a mobile application.

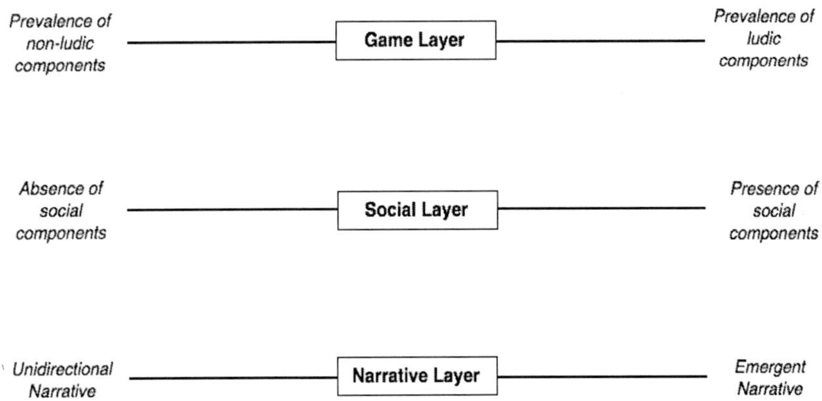

Figure 1 - The model through which we have analysed our case studies

The narrative layer aims at defining how deeply integrated in the app a narrative component is, a matter that seems to be recurrent in our case studies. Although the products we have analysed all show some sort of tension towards narrative, in our research we often found ourselves questioning the status and structure of narration within location-based applications. On the one hand, there is a tendency to encourage diverse styles of fruition (as in *Foursquare*). On the other hand, our understanding of these products as narrative objects is based on two different narrative models: that of an emerging or collaborative narrative, and that of a unidirectional, more "traditional" narrative. Though it is clear that most of the location-based ludic applications retain a narrative potential, the analysis of these peculiar media products within a narrative/narratological frame remains problematic.

In summary, our model is based on a tripartite analysis that aims at testing the ludic, social and narrative potential of the applications we chose as case studies. We argue that the objects of our research can be placed on the three proposed *continua* according to design and user experience. This three-headed model may be helpful in order to identify different tendencies and biases in location-based applications.

4. *SCNVGR*

In its official web site, *SCVNGR* is described as "a game about doing challenges at places" (Scvngr, 2011a). It is a free mobile application that supplies the player with different kinds of challenges to be undertaken in the real world. Overcoming these challenges may grant the user different benefits or virtual prizes.

In July 2010 Seth Priebatsch, founder and CEO of the game, opened his TED-Boston speech claiming that through his company's products he was "fairly determined to try and build a game layer on top of the world" (TED, 2010). Priebatsch's idea of game layer seems to adhere to the gamification's practice of using ludic mechanics to "spice up" non-game products. The use of points, rewards, badges, levels and–in some cases–of deeper ludic mechanics is certainly one of the constitutive features of this application. Along with Priebatsch's idea of a game layer, this may lead us to consider *SCVNGR* a game in a more traditional sense of the word, at least at this stage of research. The product shows deeply integrated ludic features that actively create a rule-based artificial conflict with a clear, quantifiable outcome, thus satisfying Salen and Zimmerman's definition of "game". Yet even in a perfectly clear case like that of *SCVNGR* some sort of fuzziness connotes the practices inspired by these applications. The "model user" of *SCVNGR* will typically be engaged in a pervasive ludic practice, alternating between goal-oriented behaviours ("*do this* in order to win a free pizza") and more mundane tasks. As we will see, the basic functioning of location-based applications is inherently adaptable to both traditional playing practices (e.g. *Whai Whai*, which seems to conflate treasure hunt and choose-your-own-adventure books mechanics) and fuzzier, non-teleological playful practices. Nevertheless, this distinction in the usage of the applications does not influence their ludic nature; *SCVNGR* can be placed in our model on the right end of the "game layer" spectrum. Proceeding with the analysis, we can say that the challenges of *SCVNGR* are strictly linked with the physical space they are set in, as the application involves geo-localization as a pivotal aspect. The game's claim is in fact: "Go places. Do challenges. Earn points" (Scvngr, 2011a). The challenges are created by users, by companies using them as marketing tools or by local shops and institutions that seek to promote their work and initiatives, and can be put together by *treks*–themed sets of places with dedicated challenges in each one of them–that expand the game experience. Again, from the official website, the game is described as "part awesome location-based mobile game [and] part really powerful mobile gaming platform" (Scvngr, 2011b), stimulating the production of brand new experiences after paying some money to the game owners. The narrative part of the game, then, is deeply collaborative and emergent, placing *SCVNGR* on the right side of also the "narrative layer" spectrum.

The analysis of the social layer is more complex. A key feature of the game is the fact that rewards are both "virtual" (i.e. badges and points in the game) and "real", often provided by the creators of the challenges. This duality recalls the difference between intrinsic and extrinsic work rewards described by sociologist Clifford J. Mottaz (1988, pp. 467-480). The intrinsic rewards are connected with the worker's interest in the task, they come from him. The extrinsic rewards come instead from the employee. The commitment to work depends on the proximity of these rewards with the work values (in the form of desires and expectations) the worker assigns to the job. It is clear that the values have high profile aims and, even if they are not quickly satisfied, they cannot influence commitment too much. The rewards are instead pivotal and can deeply change the worker's involvement in the job. In *SCVNGR* we face a similar situation with intrinsic–in this case not only the player's interest in the game but also in the rewards found within the game world–and extrinsic rewards. The former are bound above all to a social desire for contact, status and reputation through "virtual" prizes (badges and points); the latter are instead the "real" prizes the users can get. So, it is in the intrinsic rewards, in the social status desire, in the confrontations between users and the bottom-up productions already mentioned, that the social component of the game emerges, putting *SCVNGR* on the right side also of the "social layer" spectrum[4].

---

[4] At the end of 2011, Priebatsch launched *Level Up*, a *SCVNGR* add-on where the mixture of ludic and social dynamics was replaced by a marriage of game-like mechanics and micro-transactions. Through this application it is possible to make payments using a personal QR code, connected to the user's credit card. Goals and rewards created by sellers are based on the times and frequency of the payments rather than on the achievement of certain objectives. With *Level Up*, Priebatsch seems to put aside the narrative and social dynamics that were embedded in

5. *FOURSQUARE*

Foursquare is another mobile application that implies geolocation and urban exploration. In this case the focus is on the voluntary *check-ins* by which the users record their passages through places. Points and badges are the prizes for the users that visit the same place more often or that create and post more comments about them. *Foursquare* also produces a weekly chart–based on algorithm calculations invisible to users–that allows to elect the *mayor* of every place recorded. The approach is hereby competitive, since it's possible to lose a *mayorship* to another user.

Born with the aim to help discovering new places through a gamification scheme composed of check-ins and rewards, *Foursquare* also gains its success from its participatory nature and its two co-existent souls: the use of the product as a recording device (in the fashion of a travelogue, to share written notes on places, routes, episodes), and its use as a factitive device, similar to the use of a bottom-up guidebook, that other users can read and follow, even without contributing themselves.

Thus on the narrative level *Foursquare* changes depending on its use: as we have seen, as a travelogue it builds up a collaborative narrative, as a traditional guide book instead it remains bound to a unidirectional narration that its users choose to follow.

Also considering the social layer this differentiation remains strong. It's easy to underline that in this case narrative and social components are deeply linked. Collaborative narration assigns a pivotal role to the social component, in the form of shared comments, pictures, activities, tips and To-Dos that lead other users' experience. Even badges and mayorship competition, tracked through dynamic leaderboards, are vital social aspects of *Foursquare*. Another important feature, common to all social media, is the parasitical gathering of new users (as friends or competitors) from other major social networking websites through invitations. So, while in the first case the social aspects are vital to the *Foursquare* experience, where the narration is unidirectional this social layer is thinner and the application becomes a sort of bottom-up guide book.

SCVNGR, using the game layer simply as a tool to encourage commercial transactions. Ludic components are deeply dampened and the process proposed by *Level Up* seems to be akin to accumulating fidelity points in chain stores.

Analysing the game layer produces less predictable results compared to those from the previous case studies. Here, the application is deeply permeable and allows users to move freely from an openly ludic style to a more utilitarian one. Rewards, points, rankings, mayorships and badges may be good examples of this dynamics, as such features act as motivational catalysts, stimulating "competitive" exploration. This practice is not just directed to the simple discovery of new places to visit–in most cases suggested by a network of friends–but driven by hoarding, prize-hunting and competitive dynamics[5]. This duality prevents us from placing *Foursquare* on the far right end of the "game layer" spectrum. As a matter of fact, using the application as a guidebook will produce a poor ludic interaction–or even no gameplay at all–while *Foursquare*'s system of check ins and rewards, when used as a travelogue, will point the user towards a semi-ludic experience. To account for this fuzziness and for the absence of clearly defined ludic challenges–similar to those found in *SCVNGR*–we decided to place *Foursquare* in a more nuanced position.

6. *BROADCASTR*

Broadcastr was launched in December 2010 by Andy Hunter and Scott Lindenbaum, who described it as "an app for iPhone and Android that creates intimate and immersive experiences by unlocking pictures and audio relevant to where you are. It turns your smartphone into a multimedia guide to the world, and everyone can contribute" (Broadcastr, 2010). *Broadcastr* is based on the user's production of micro-narrations that are connected to places found in the real world. Typically, a *Broadcastr* user will record his story about a particular spot in a city through the voice recording plug-in included in the application. This audio fragment is then made available to all *Broadcastr* users who will be able to access it from their mobiles or home computers. Unlike *Foursquare* or *SCVNGR*, *Broadcastr* does not require its users to be physically present in a place in order to be able to browse and contribute to the audio library of the application. *Broadcastr* can be used from home and the audio files are all accessible at once from any location on the planet. Nevertheless, in their statement regarding *Broadcastr*, Hunter and Lindenbaum seem to encourage a geo-localized use of their application. On *Broadcastr*'s website the two founders write:

---

[5] A more articulate analysis of the usages of location-based digital games can be found in Souza e Silva & Hjorth, 2009.

> Go exploring with *Broadcastr* and you'll find memories, insights, and enriching information about eclectic and everyday places on every continent on Earth. Take a walk while stories about your surroundings stream automatically to your phone. A celebrity chef whispers in your ear as you stroll past his favorite restaurant; a renowned architect guides you through lower Manhattan; a comedian shares a hilarious personal anecdote at her favorite bar. Your movement through the world becomes your search query. Download the app. Take a walk (Broadcastr, 2011).

Therefore, while it is possible to browse *Broadcastr* remotely, using it on a smartphone while actually moving across the city will create a playlist of audio fragments that will be played as the user reaches a certain spot. In this case we may find another example of divergent affordances or possible usages of a location-based application. *Broadcastr* can be used both as a random access tour guide from a home computer and as a geo-related playlist creator supplying a flux of micro-narratives to a user traversing a real space. This makes the narrative component of *Broadcastr* strikingly similar to the one found in *Foursquare*. In both cases, an emergent location-related narration is built upon a multitude of user-created fragments, but the application allows a less enterprising user to resort to *Broadcastr*'s flux as one would do with an audio guide. Nevertheless, in this case, the application cannot be compared to a proper guide book, due to the fragmented nature of the contributions.

Of the four case studies we focused on, *Broadcastr* is the only one that avoids showing clearly definable ludic features. The application does not include badges or a rewards system and poses no explicit challenges. Nevertheless, the openness of the system makes it a potential platform for playful practices designed by the users. One could easily imagine a scavenger hunt game where hints are administered through audio fragments that are activated when players reach a certain spot.

Broadcastr's dynamics of social interaction seem less prominent than those in *Foursquare* or *SCVNGR*. While it is possible to *follow* friends on *Broadcastr* (much like what happens on *Twitter*) and comment on their profiles, the discreteness and diversity of users' contributions, ranging from journal entries to field recordings, and the absence of any defined competitive structure (e.g. mayorships in *Foursquare*) make user interaction much looser and rarefied.

7. *WHAI WHAI*

Whai Whai is a series of games challenging players to explore the central areas of a small number of cities by asking questions that can be answered only by actually going to a specific place to observe the details of a building or a landmark. Each *Whai Whai* game covers a specific city (Florence, Rome, Milan, Venice, etc.). It aims at making tourism more compelling and engaging by assigning players a series of quests and riddles to be solved, tracking their progress throughout the game and adding a developing narrative unfolding across each session. *Whai Whai* can be played using an iPhone or a specific paper booklet and an ordinary mobile phone and, obviously, each episode must be played in the corresponding city. Users progress through a narrative designed for a specific urban area by exploring it, examining its landmarks and answering the questions proposed by the game.

This application is much more closed than the previous examples. Differently from the other cases, *Whai Whai* games are designed from a single team that does not leave to ordinary users the power to alter the game structure and to add new contents. While its branching structure is remarkably wide and makes it possible for subsequent games to be quite different from one another, it is still a finite experience that cannot be expanded.

Also, game-like elements are more evident in *Whai Whai*. Sessions have definite beginnings, developments and endings, there is a rudimentary score system and the game evaluates the players' progress. While the system is designed not to allow users to *lose* a game–the story and the session will progress anyway even if the wrong answers are repeatedly given–it still signals whether players are performing in a good or bad way.

Whai Whai, then, is designed to encourage a specific type of movement in urban areas, closely related to the practices of visiting and exploring a city. Although it may certainly be possible to witness emergent behaviours during a *Whai Whai* game, those are not openly invited by the system due to the closed and unidirectional (top-down) nature of its ludic and narrative components.

Conclusions

Through a tripartite model (Figure 2) we tried to analyse our four cases basing our assumptions on both their design and the practices they elicit. In two cases (*Foursquare* and *Broadcastr*) this approach led us to the decision of considering two different modes of engagement that we defined as guide book and travelogue. For this reason, the two applications are not placed univocally on the *continua*, but are split into two distinct entities (b1 and b2 for *Foursquare*, d1 and d2 for *Broadcastr*) occupying different positions in the scheme. Our distinction is based on the discrete styles of usage that emerged from the analysis of the affordances of each application. We must stress that the usages that we propose are theoretical; we are aware that the empirical user of *Foursquare* or *Broadcastr* will enact hybrid ludic practices, where clear-cut distinctions cannot be made.

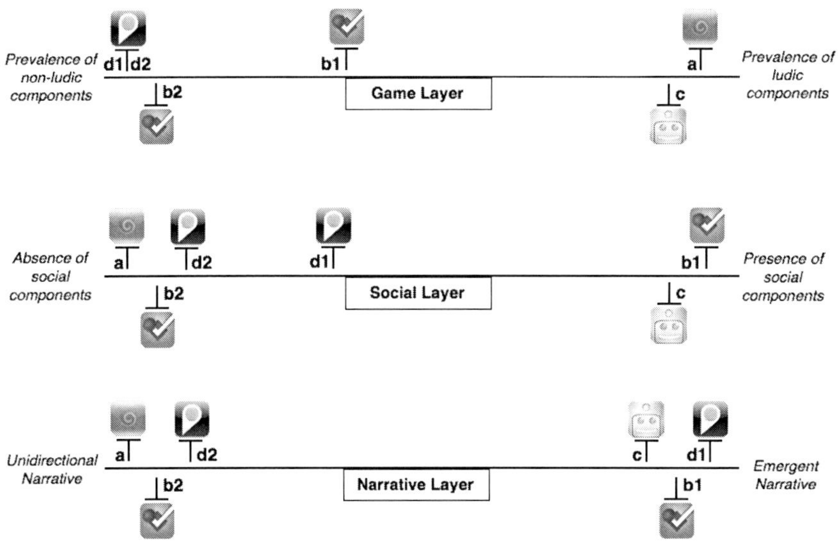

Figure 2 - The complete model through which we have analysed our case studies. Legend: a-Whai Whai; b1-Foursquare (travelogue use); b2-Foursquare (guide book use); c-SCVNGR; d1-Broadcastr (travelogue use); d2-Broadcastr (guide book use).

While the four applications that we decided to analyse seem to rely on common premises (turning everyday spatial practice into *something else*), our model shows the complex and varied nature of these products.

SCVNGR and *Whai Whai* seem to take different approaches to the ludicisation of urban exploration. *SCVNGR* is consistently featured on the right side of the model. This means that a strong social component and an emergent approach to narrative complement a clearly defined ludic orientation. Of our four specimens *SCVNGR* is the one that adopts the most pervasive approach, where playful practices are intertwined with social networking and contribute to the construction of a collaborative narration. On the other hand, *Whai Whai* relies on a consistent game apparatus with defined rules and goals but does not allow its players to access to their expanded network, nor to engage in an activity of collaborative narrative construction. In this sense, *Whai Whai* can be seen as a "ludicisation" of an older medium such as the guide book or, on the other hand, as a spatialisation of hypertextual media such as choose-your-own-adventure books. Foursquare poses different problems since two distinct possible usages emerged from the analysis. If used as a user-generated guide book (case b2), the application is consistently featured on the left side of the scheme, where less interaction with the user occurs. In the case of what we defined the travelogue usage (case b1), Foursquare seems to mimic SCVNGR's behaviour, since it features a strong social component and a predominantly collaborative narrative development. What stands out in this case is *Foursquare*'s lack of a distinct ludic component. While *SCVNGR* and *Whai Whai* aim at creating a *gaming situation*[6], where rules and goals can be observed, *Foursquare*'s use of game-like features (rewards, points and rankings) does not offer proper challenges for the user[7], but merely rewards trivial actions such as visiting a museum or dining in a restaurant. For this reason, on the game layer, *Foursquare* is placed in a hybrid position.

Broadcastr was mostly used as a sort of *contrast medium*, since no clear ludic components could be observed. Because of its scarcely factitive nature (it does not propose challenges nor explicitly elicits playful practices), *Broadcastr* stands out as a platform for different modes of

[6] Finnish researcher Markku Eskelinen defined the gaming situation as "a combination of ends, means, rules, equipment and manipulative actions" (2001).

[7] Practices as FAQs, walkthroughs and even cheating are also quite common among *Foursquare* users. It is easy to find web sites and online communities explaining how to unlock special badges or how to exploit the inconsistencies of the software to obtain special outcomes without physically being in the place requested (e.g. Royal Wedding Badge). Even if clearly inscribed in the area of unpredicted uses, all these cases contribute to posit *Foursquare* into a peculiar ludic frame.

expression. It is not surprising then, that it hosts a wide variety of contributions ranging from the autobiographical to the fictional. On the narrative continuum, *Broadcastr* replicates *Foursquare*'s usage-based distinction, since it can be used as a guide book, retaining unidirectional narrative dynamics, or as an audio travelogue. Also on the social layer *Broadcastr* reproduces *Foursquare*'s features, but, as we have seen in the application's analysis, with less prominence of the social networking capabilities.

At this stage of our research, we cannot formulate a conclusive hypothesis on the common traits of location-based applications yet. However, the multi-layered, multi-dimensional analytical model has proven flexible enough to adequately produce relevant and meaningful descriptions for our case studies. We believe that this preliminary result is worth underlying and future research will proceed in the same direction. As temporary closing remarks, we want to stress the usefulness of building heuristic models based on both the features of the case studies and their potential usages when dealing with complex artefacts such as those analysed here. For instance, in this short comparative analysis, we built a multi-dimensional model by deconstructing the term "location-based" and the critical discussion of the term "gamification" in order to provide a satisfying description of the various features of our case studies.

References

Broadcastr, (2010). *About* [web] Available at: http://broadcastr.com/betablog/about/ [Accessed 11 January 2012].

Caillois, R. (1991). *Les jeux et les hommes: la masque et le vertige* (4th ed.). Paris: Gallimard.

Eskelinen, M. (2001). The Gaming Situation. *Game Studies*, *1* (1). Available at http://www.gamestudies.org/0101/eskelinen/ [Accessed 11 January 2012].

Goffman, E. (1961). *Encounters. Two studies in the sociology of interaction*. Basingstoke: MacMillan.

Greimas, A. J. (1983). *Du sens. 2*. Paris: Éditions du Seuil.

Huizinga, J. (1955). *Homo Ludens: A Study of the Play-Element in Culture*. Boston: The Beacon Press.

Montola, M., Stenros, J., & Waern, A. (2009). *Pervasive Games. Theory and Design*. Burlington: Morgan Kaufmann Publishers.

Moore, C. (2011). The Magic Circle and the Mobility of Play. *Convergence*, *17* (4), p. 373-387.

Mottaz, C. J. (1988). Determinants of Organizational Commitment. *Human Relations*, *41* (6), p. 467-480.
Prell, C. (2011). *Social Networks Analysis: History, Theory and Methodology*. Thousand Oaks: SAGE Publications.
Salen, K., & Zimmerman, E. (2004). *Rules of Play: Game Design Fundamentals*. Cambridge: MIT Press.
SCVNGR, (2011a). *Homepage* [web] Available at: http://www.scvngr.com/ [Accessed 11 January 2012].
—. (2011b). *Build on SCVNGR* [web] Available at: http://www.scvngr.com/builder [Accessed 11 January 2012].
Souza e Silva, A., & Hjorth, L. (2009). Playful Urban Spaces: A Historical Approach to Mobile Games. *Simulation Gaming*, *40* (5), 602-625.
TED (2010), *Seth Priebatsch: The Game Layer on Top of the World* [video] Available at: http://www.ted.com/talks/seth_priebatsch_the_game_layer_on_top_of_the_world.html [Accessed 11 January 2012].
Virrantaus, K., Markkula, J., Garmash, A., Terziyan, V., Veijalainen, J., Katanosov, A., Tirri, H. (2002), *Developing GIS-supported Location-Based Services* [online]. Available at: http://www.cs.jyu.fi/ai/papers/WGIS-01.pdf [Accessed 11 January 2011].

CHAPTER XI

CITIZEN JOURNALISM AFTER A NATURAL CATASTROPHE: THE EMERGENCE OF AN ALTERNATIVE PUBLIC SPHERE

MANUELA FARINOSI AND EMILIANO TRERÉ

1. The L'Aquila earthquake

In 2009 a 6.3 magnitude earthquake struck L'Aquila, a thirteenth-century city on the mountains of central Italy. The seism caused serious damage to L'Aquila and the surrounding villages, destroying many parts of the medieval centre, killing more than 300 people and leaving 65,000 people homeless. Immediately after the disaster, numerous citizens of L'Aquila started to use Internet platforms on a massive scale in order to voice their opinions and share information related to the actual situation people were undergoing during the post-earthquake phase (Farinosi & Micalizzi, 2012). These online environments offered citizens new channels for reporting, speaking and acting together and substantially contributed to an explosion of citizen journalism practices. An array of social media platforms, blogs and content-sharing sites was flooded with news, posts, comments, videos and photos related to issues regarding the emergency and the post-emergency situation, daily life in L'Aquila after the devastating earthquake, the city's rebuilding problems, the recovery efforts and the social re-appropriation of public spaces damaged by the seism. This work explores the production of grassroots information in a context that could be described as "out of the ordinary". We first provide a review of current literature on the phenomenon of citizen journalism (section 2). Then we illustrate the aims and methods adopted (section 3) and describe and discuss the findings of our research (section 4). Finally we draw some conclusions and sketch some future challenges for research into this topic (section 5).

2. The changing landscape of the press: digital media and citizen journalism

From the early years of the 21st century, society has been infiltrated by new digital technologies with potentially profound consequences and implications for the power dynamics of participation among citizens. The Internet has become an integral part of people's life and nowadays citizens increasingly voice their opinions on the social web. Web 2.0 platforms exhibit new possibilities and opportunities for mobilization, organisation and discussion, offering ordinary people the chance to create and spread online news, information and commentary. It has also been demonstrated that the use of 2.0 applications in response to a disaster can contribute to building resilience in the community affected by the catastrophe (Schmidt, 2010; Bruns et al., 2012; Dufty, 2012; Taylor, 2012). The new grassroots practices which have developed around digital media also contribute to the modification of established media organisations. Traditional news media and professional journalists have undergone significant changes as numerous motivated amateurs around the world have begun to use Web 2.0 applications to spread news on specific and localized topics and to document important social issues. The online publication of user-generated, real-time content is altering the way in which news are produced and diffused, and on several occasions, online hyper-local initiatives have depicted different versions of the reality of particular stories (Schaffer, 2007). Scholars and professionals have addressed this new phenomenon by adopting a wide array of labels, and it has no agreed definition, although the term "citizen journalism" is now generally accepted (Littau, 2007; Carpentier, 2008; Allan, 2009; Atton, 2009; Goode, 2009; Lewis et al., 2010). Even if a review of the academic and professional literature yields a variety of different applications for this kind of media practice, they all imply the participation of a group of citizens, without professional journalistic training, in the process of reporting, collecting, sharing and disseminating news and information (Bowman & Willis, 2003; Lasica, 2003; Gillmor, 2004; Glaser, 2006; Jarvis, 2006). According to Metzgar et al., the production of independent information–often community-oriented and linked to locally based issues–is "intended to fill perceived gaps in coverage of an issue or region and to promote civic engagement" (2011, p. 774). Mark Deuze argues that citizen journalism marks a positive revival of "civic, communitarian, or public journalism" (2009, p. 256). Patricia Ho describes this phenomenon as "a return to the fundamental goal of representing society's voice in the news" (Ho, 2011, p. 42) and several authors assign values such as truth-telling, impartiality and unbiased observation to the work of amateur journalists. To engage community members in the reporting process means, in fact, that news stories are no longer only revealed at the discretion of the traditional mainstream media but, due to the desires of the community itself, information is transmitted through "human filters", so that useful or interesting information is reposted and shared online. Different scholars agree to some extent that at the core of this phenomenon lies news-content which is produced every day by unpaid people and uploaded on independent news sites, social media and file-sharing platforms like YouTube or YouReporter. Nowadays, in fact, thanks to the diffusion of digital media, communities can reveal the effective reality of their daily life through their own words, and internet-based media have been redefining the public sphere and enlarging the democratic space of expression all over the world.

At the international level, important examples of grassroots journalism include the following online portal and news sites: Digg.com, Newsvine.com, Globalvoicesonline.org and Nowpublic.com. In the Italian context, significant examples are represented by Internet news sites such as Linkiesta.it, Agoravox.it, Cafebabel.it, Lamianotizia.com and Fainotizia.it. Repubblica.it, the website of one of the most powerful Italian daily newspapers, recently launched a special section dedicated to citizen journalism[1]. This initiative is part of a wider project[2] started in July 2011 aimed at investigating topics proposed by the newspaper's audience such as mafia issues, corruption in political elites, environmental disasters and healthcare problems. Moreover, in the last few years the Italian micro Web TV's ecosystem has also played an increasing relevant role in monitoring and reporting several topics not covered by the mainstream media (Treré & Bazzarin, 2011). All the user-generated content uploaded on these sites is community-focused and provides some coverage of important community issues. According to Goode, it is possible to define citizen journalism as a:

> Complex and layered mix of representation, interpretation (and re-interpretation), translation, and, indeed, remediation (Bolter & Grusin,

---

[1] inchieste.repubblica.it/it/repubblica/repit/2011/09/29/news/il_cittadino_si_ribella-22447084/

[2] inchieste.repubblica.it/

> 1999) whereby news and comment, discourse and information, is reshaped as it traverses a range of sites and varying media platforms (Goode, 2009, p. 1291).

It is important to underline that, even if citizen journalism is generally associated with the Internet, there are several crossovers between conventional media and citizen journalism (Goode, 2009). Here, we share Goode's definition and consider grassroots news production as an intricate phenomenon composed of a variety of news-making practices using multiple applications (blogs and micro-blogging platforms, social media, alternative and/or independent websites, mobile phones, digital cameras and other portable devices). To conceive of citizen journalism as a continuum of integrated practices means primarily that the focus of our research is not on the medium, but on the people.

The example of the L'Aquila post-earthquake media landscape is illuminating in this respect: while citizens created their own blogs and websites to provide their point of view on the situation of the city, at the same time they massively appropriated corporate online applications like Facebook and YouTube to spread the content they produced. Using multiple websites, personal blogs, social networking sites and video/photo sharing platforms, many ordinary citizens of L'Aquila were able to post articles, reports, videos and pictures in order to cast light on the post-quake situation in great detail, illustrating many of its nuances, providing their own reflections on the tragedy, and narrating their everyday lives from the inside (Farinosi & Treré, 2010; Farinosi & Treré, 2011; Padovani, 2010).

Citizens from L'Aquila used the Internet to bypass traditional gatekeepers (Bennett, 2003; Rucht, 2004) and communicate their critical views on their post-quake situation directly. Moreover, activists, collectives and civic movements (such as, for instance, the movement of the "People of the Wheelbarrows" and the "3e32" collective) employed multiple Internet technologies including Web 2.0 platforms like Facebook, YouTube, Twitter, Flickr, Web TVs and blogs to organise and report citizens' protests and to provide their perspectives on the post-quake events in L'Aquila (Padovani, 2010; Farinosi & Treré, 2010; Farinosi, 2011; Treré & Farinosi, 2012).

3. Aims, conceptual framework and methods

In order to investigate citizen journalism practices in the post-quake L'Aquila situation, we deployed a qualitative research method. We selected twenty citizens amongst the most active online creators of bottom up information and in April 2010 we carried out twenty in-depth interviews. Among these active citizens were bloggers, amateur journalists and video-makers, who were either residents of L'Aquila or otherwise interested in the L'Aquila post-earthquake situation. Most of them–as well as most of the people engaged in the production of grassroots information in L'Aquila–were educationally very qualified. Some were doctors, university professors and graduate students and some had their backgrounds in Computer Science or Communication Sciences. The interviews ranged from 45 to 70 minutes in length and aimed at exploring the motivations for producing and publishing grassroots content online. Each interview was recorded and transcribed verbatim and the resulting texts were analysed for categories that signalled the citizens' perspectives on the grassroots productions.

The data analysis was approached without any preset list of categories and understanding emerged through interaction with the available data. The interviews were analysed by using the theoretical framework of Social Action Theory together with the Uses and Gratifications Theory. According to Weber's definition, social action is "such action as, according to its subjective meaning to the actor or actors, involves the attitudes and actions of others and is oriented to them in its course" (Weber, 1907 [1956], p. 3). The application of the Uses and Gratifications Theory allows for the exploration of the different categories of needs fulfilled by the use of online platforms, such as Facebook, YouTube and blogs. The combination of both models and the application of a qualitative approach created an understanding of the phenomena from the perspective of the population under study (Vettehen et al., 1996). The adoption of this framework led to a clearer view of the motivations for human behaviour and a better understanding of the actors' points of view, motivations, goals and values.

Furthermore, our study adds valuable qualitative data to the already existing literature regarding citizen journalism, which is mainly quantitative in nature (Korgaonkar & Wolin, 1999; Sheehan, 2002; Li, 2005). In fact, while scholars have already analysed the production of grassroots information from numerous perspectives, only a few have explored the question of why people do what they do (see for instance Trammel et al., 2004; Nardi et al., 2004; Ekdale et al., 2010; Kokenge, 2010).

4. Findings

In this chapter we focus on the two most important factors that pushed L'Aquila citizens to produce grassroots journalism in the post-earthquake scenario: 1) providing an alternative view to the mainstream representation of the disaster and 2) rebuilding broken ties through digital media.

4.1. An alternative view of the mainstream representation of the disaster

The primary reasons for producing citizen journalism originated from the fact that the mainstream media coverage of the post-earthquake situation in L'Aquila was perceived by citizens as twisted, partisan and overabundant, especially in the first days after the catastrophe. In response to such biased media representation, people in L'Aquila were motivated to provide their own version of the facts through web platforms. Commenting on the information produced by mainstream media, Francesco, a young Communications student and amateur video maker, spoke about what he called "information bulimia" and pointed out the propagandist aspect of "information that often winked to the Italian government".

Anna explained to us that she turned into an amateur reporter because "the main problem of this earthquake [...] was the information, namely the lack of information at all levels". According to Ezio, another active blogger, after watching the news transmitted by Italian TV, people living outside L'Aquila felt that "everything was ok":

> Everything is ok in L'Aquila. This was the message from the beginning [...] My son lives in Milan and came to L'Aquila for the Easter holidays. He came and I heard from him this thing that went on television: L'Aquila's all right...I haven't heard it before! What does this mean? I could not even understand the meaning of "L'Aquila was ok" and then he gave this title to a video he had done, "In L'Aquila everything is okay", in an ironic tone, then I began to realize that actually outside L'Aquila there was the idea that here everything was going right, according to an ideal "Aquilano miracle" that the media were mentioning in the news.

Various other studies have underlined how the Italian mainstream media first framed the earthquake as a spectacle of catastrophe (Treré & Farinosi, 2012) and then depicted the post-quake situation in a propagandist manner as a "miraculous reconstruction" by the Berlusconi government (Padovani, 2010; Ciccozzi, 2010). In this regard, the interviewees also cited some example of news transmitted by the mainstream media in the days after the earthquake. Several citizens told us about the case of a City Council meeting held among the rubble in the centre of L'Aquila and the fact that the TG5 newscast (the most important channel of the Mediaset group, owned by the former Italian Prime Minister, Silvio Berlusconi) twisted the news by reporting that a council meeting was being held outdoors to celebrate the reopening of some of the city centre's most important shops. In reality, the mayor of L'Aquila had decided to summon the council in the main square in order to protest against the critical post-earthquake situation of the city. Federico, a young blogger, explained that:

> While the TG5 said that the City Council had met on the ruins of the Piazza Palazzo to celebrate the reopening of the historic centre, in fact there was a press release of the municipality stating that "we join to protest against the fake media reconstruction of the old town".

Federico also points to an emblematic episode experienced by his grandmother that can be used to explain some journalists' desire to twist reality by showing only a small portion of the truth about the earthquake. Federico's grandmother was interviewed by a TG5 journalist, who then cut most of the speech and left only the part where she mentioned her happiness about the centre coming back to life. It is worth reporting this in Federico's words:

> A news report by TG5 made me mad! Here the Bar Nurzia in Piazza Duomo had reopened, and they had the audacity to come into town and do their shooting. They also interviewed my grandmother, among others. On the day of reopening, the bar Nurzia produced its famous nougat of L'Aquila, but they were practically the only ones to open! My grandmother went to the centre immediately because she was so happy that they were open, and also because Christmas was approaching and she wanted to buy presents for her relatives. Well, she was interviewed, and they asked "Madam, are you happy because the centre has reopened?" And my grandmother replied, "Does it look like the centre has reopened? This is a list of condolences that we are doing! A shop here, another there, the course is underpinned, is closed and there are soldiers to guard the 'red zone' […] if you think the centre has reopened [...] we come here because

> this is the only thing that has reopened and we are trying to return to a normal life [...]" But when I saw the service on TV it was like: "Aquilani are delighted that the centre has completely reopened." The interview with my grandma was cut without any shame, to give deliberately false news.

Another example that emerged from our analysis described how mainstream media neglected the so-called "20,000 protest" of June 16th, 2010. On that occasion, the news was reported only by few local TV stations, while the most important Italian channels did not mention the event at all.

All the cases we reported generated a deep frustration and this constituted the main engine for producing grassroots information. Mainstream media coverage was perceived as depicting what was far from the real situation of the city, therefore citizens from L'Aquila started to produce and publish their own online information about their daily realities, in order to give voice to their points of view, to provide their perspective on the catastrophe and thus create a contrast with the messages spread by the national media. Francesco, an avid video-maker, stressed the importance played by grassroots information in documenting real life:

> I believe strongly in citizen journalism, in this powerful tool that people, citizens, use to communicate and carry out delivering information and, above all, to tell their stories. The Internet has proven to be an important tool to tell our reality and to share it with other people, firsthand, from below, straight from our eyes, in the city.

Telling stories through digital media applications constitutes a strong way to capture and reflect on certain traumatic events, and it can also help people who are in the process of recovering from the loss of their daily routines. The skill to capture the moments was also fundamental in the first hours after the main shock, when the cameras of national and international televisions were still absent. This aspect was highlighted by Thomas in detail:

> The role of user generated content was pivotal because you could not tell the story of the earthquake without images of people who had suffered these things first hand and who had the courage to go there to report the events in the hours immediately following the earthquake. We have a lot of survivors who carried out such actions and thanks to them we are now able to remember and live again those moments [...] The only way to tell them is to live them really; this is the strength of today's technology!

4.2. Rebuilding broken ties through online public spaces

Another strong reason for producing grassroots information is related to having the will to share the content created within a community. The material that was posted online by amateur journalists generated several discussions with other citizens in relation to the rebuilding process and other important issues concerning the post-earthquake phase. During the emergency, many people were forced to leave their houses and move to tent cities or to other Italian locations. Given the impossibility of interacting or meeting on a daily basis, these people tried to rebuild the ties that were cut off by the tragedy using digital technologies in order to reconstruct their lost public spaces online. Furthermore, these online audiences were also useful in motivating amateur journalists to continue their alternative news production. The idea is "to rebuild the social fabric that has been lost, and to rebuild it according to more ethical principles. In this respect, the Internet can play a fundamental role" (Alessio, activist). As blogger Anna stated: "it's important to let people outside L'Aquila know the situation, but it's also fundamental to create an online space for local citizens to join together". According to many citizens, online applications were spreading a different kind of content in opposition to conventional media, an "information with a heart […], because obviously being involved in this we can put our emotions and we can communicate our feelings" (Luca, videomaker).

A very interesting aspect that emerged from the interviews is that, while before the earthquake many people did not have an account on any social platform and conceived these applications as nothing more than "silly games for adolescents" (in the words of Giovanni), the situation radically changed after the catastrophe, when several citizens began to join social media to create and spread alternative information and to build connections and share emotions with other people.

Enza owned a blog before the tragedy and used it in her spare time, but did not have a Facebook account. She pointed out during the interview that she used to conceive this platform as just a waste of time, not as a social resource. After the earthquake, however, her vision changed. She "wasn't alone anymore" and Facebook represented "the first thing to do" in order to "maintain contacts with friends and to know where they are, what they are doing" thus to regain contact with her community, in particular with lost friends and neighbours.

The similar case of Giovanni is also paradigmatic in this respect. Before the earthquake, he did not have a Facebook account and thought about social media as a waste of precious time. But after the tragedy, things radically changed when he decided to open his Facebook profile. He then realized that:

> In a few days an amazing thing happened: many people started to answer my requests for friendships and contacted me. I found people who I hadn't seen in ages and I then realized the power of this social networking phenomenon. As I said, I did not believe in this phenomenon of social networking. For me it was just a series of lines of computer code, I just could not see the look, the impact of this thing. It was just a game for me, but then you cannot imagine the social impact that a thing like that could have.

Moreover, Giovanni decided to open his own Facebook page: "Aggregatore Aquilano di Flussi Alternativi"[3]–where he collected and then spread alternative and/or underreported information created by local journalists, acting as online independent "news-catalyst". On the one hand, many people visited his Facebook page to acquire a more nuanced vision of what was happening in L'Aquila. On the other hand, citizens also used this online space to build connections, start discussions and share their feelings about the reconstruction process.

5. Final remarks: the emergence of an alternative public sphere

It emerged from our findings that digital media were used to give more visibility to under-reported events and to shed light on the daily struggles of citizens that had been neglected by the Italian conventional media. Also, they worked as platforms to re-establish connections among a dispersed and traumatized community who was looking to rebuild a social dimension that had been seriously compromised because of the earthquake.

Nowadays, according to some scholars, the horizontal flow of self-made information is undermining the structures of the centralized political order, providing ongoing opportunities for dialogue and discussion and

[3] https://www.facebook.com/pages/Aggregatore-Aquilano-Flussi-Alternativi/181281390915

facilitating the formation of discursive networks which offer alternative perspectives on both national and international issues (Ford & Gil, 2001; Downing, 2003; Lievrouw, 2011). Fraser (1993) has used the term "counter-public sphere" to denote alternative discursive arenas where subordinated groups can deliberate, articulate and circulate oppositional interpretations of their own identities, interests, needs, strategies and objectives. In reference to the totality of alternative media, Fuchs (2010) prefers to talk about an alternative public sphere, a sphere of protest and political discussion that enhances the vividness of democracy.

Thus, the results of our research highlight the pivotal role played by the online media in the emergence and construction of an alternative public sphere in the post-earthquake scenario. We have seen that local citizens used digital media in order to provide an alternative version to the one displayed by the Italian mainstream media. Through the use of personal blogs, social media and sharing applications, citizens of L'Aquila were able to cast light on their "out of the ordinary" lives in the fullness of its aspects and nuances. Feeling marginalized by the mainstream media, amateur journalists bypassed traditional gatekeepers (Bennett, 2003; Rucht, 2004) and directly communicated their points of view about the post-quake situation, while at the same time strengthening their community. The totality of grassroots information (comments, posts, tags, pictures, videos, audio files, text documents, Facebook groups and pages, etc.) created, uploaded and shared on the Internet regarding the post-quake scenario in the city of L'Aquila therefore constituted an alternative public sphere, which was able to provide multiple point of views about the city from a bottom-up perspective.

Our research, however, had some limitations: first of all, the situation of L'Aquila, a city in a state of emergency, was so out of the ordinary that the results are in no way generalisable to the entire population of citizen journalists. Furthermore, the Italian media and political scenario, within which we witnessed the explosion of this grassroots phenomenon, presents several anomalies that do not allow for simple comparisons with other similar cases (Mazzoleni et al., 2011).

Nonetheless, the exploration of this alternative public sphere in this particular scenario can help to shed light on the complex phenomenon of citizen journalism and on the multiple grassroots practices by which ordinary citizens tell their stories, share their points of view and communicate their emotions in a situation where the conventional media are not able to answer the population's needs for balanced and truthful information. While we do not argue that grassroots production has significantly eroded the power of the Italian mainstream media, we have to point out–as the L'Aquila case clearly demonstrates–that digital technologies can provide citizens with powerful tools for the production of independent, alternative and more democratic information. When traditional media coverage fails, people can turn into journalists and use online platforms to spread news from the inside, according to the needs of their community.

References

Allan, S. (2009). Histories of citizen journalism. In E. Thorsen (Ed.), *Citizen Journalism: Global Perspectives* (pp. 17-31). Berlin: Peter Lang.

Atton, C. (2009). Alternative and citizen journalism. In K. Wahl-Jorgensen & T. Hanitzsch (Eds.), *The Handbook of Journalism Studies* (pp. 265-278). New York: Routledge.

Bennett, W. (2003). New Media Power: The Internet and Global Activism. In N. Couldry, & J. Curran (Eds.), *Contesting Media Power: Alternative Media in a Networked World* (pp. 17-37). Maryland: Rowman and Littlefield.

Bolter, J. D., & Grusin, R. (1999). *Remediation: Understanding New Media.* Cambridge: MIT Press.

Bowman, S., & Willis, C. (2003). *We Media: How Audiences are Shaping the Future of News and Information.* Reston: The Media Centre at the American Press Institute. Retrieved from http://www.hypergene.net/wemedia/download/we_media.pdf

Bruns, A., Burgess, J., Crawford, K.. & Shaw, F. (2012). *Qldfloods and @Qpsmedia: Crisis Communication on Twitter in the 2011 South East Queensland Floods* (Research Report). Brisbane: ARC Centre Of Excellence For Creative Industries And Innovation. Retrieved from http://cci.edu.au/floodsreport.pdf

Carpentier, S. (2008). How Online Citizen Journalism Publications and Online Newspapers Utilize the Objectivity Standard and Rely on External Sources. *Journalism & Mass Communication Quarterly,* 85(3), 531-548.

Ciccozzi, A. (2010). Aiuti e Miracoli ai Margini del Terremoto de L'Aquila. *Meridiana: rivista quadrimestrale dell'Istituto Meridionale di Storia e Scienze Sociali,* 65, 227-255.

Deuze, M. (2009). The Future of Citizen Journalism. In S. Allan & E. Thorsen (Eds.), *Citizen Journalism: Global Perspectives* (pp. 255-264). Berlin: Peter Lang.

Downing, J. (2003). Audiences and Readers of Alternative Media: The Absent Lure of the Virtually Unknown. *Media, Culture & Society*, 25(5), 625-645.

Dufty, N. (2012). Using Social Media to Build Community Disaster Resilience. The *Australian Journal of Emergency Management*, 27(1), 40-45.

Ekdale, B.K., Nasmkoong, K., Funger, T.K. & Perlmutter, D. (2010). Why blog? (then and now): Exploring the Motivations for Blogging by Popular American Political Bloggers. *New Media & Society,* 12(2), 217-234.

Farinosi, M. (2011). *Civic Engagement and Community Empowerment After a Natural Disaster: The Case of the '3e32' Citizens' Committee.* In L. Stillman (ed.), *CIRN Conference Proceedings: "To Measure or Not to Measure: That is the Question"* - Monash University.

Farinosi, M., & Micalizzi, A. (2012). L'Aquila 2.0: Partecipazione dal Basso nel Primo Disastro Italiano dell'era Digitale. In E. Minardi & R. Salvatore (Eds.), *O.R.eS.Te. Osservare, Comprendere, Progettare per Ricostruire a Partire dal Terremoto dell'Aquila*, (pp. 157-176). Faenza: Homeless Book.

Farinosi, M., & Treré, E. (2010). Inside the "People of the Wheelbarrows": Participation Between Online and Offline Dimensions in the Post-Quake Social Movement. *The Journal of Community Informatics,* 6(3). Retrieved from http://ci-journal.net/index.php/ciej/article/view/761/639.

Farinosi, M., & Treré, E. (2011). Between the Square and the Net: Civic Engagement and Participation in a Post-Earthquake Movement. Paper presented at 'A Decade in Internet Time': Symposium on the Dynamics of the Internet and Society, Oxford Internet Institute, Oxford, 21-24 Settembre 2011. Available from *SSRN eLibrary* http://papers.ssrn.com/sol3/papers.cfm?abstract_id=1925852

Ford, T., & Gil, G. (2001). Radical Internet Use. In J. Downing (Ed.) *Radical Media: Rebellious Communication and Social Movements,* (pp. 201-234). Thousand Oaks: Sage.

Fraser, N. (1993). Rethinking the Public Sphere: A Contribution to the Critique of Actually Existing Democracy. *Social Text*, 25/26, 56-80.

Fuchs, C. (2010). Alternative Media as Critical Media. *European Journal of Social Theory,* 13(2), 173-192.

Gillmor, D. (2004). *We the media: Grassroots Journalism by the People, for the People*. O'Reilly Media, Sebastopol, 2004.
Glaser, M. (2006, September 27). Your Guide to Citizen Journalism. Message posted to http://www.pbs.org/mediashift/2006/09/your-guide-to-citizen-journalism270.html
Goode, L. (2009). Social News, Citizen Journalism and Democracy. *New Media & Society* 11(8), 1287-1305.
Ho, P. (2011). *Citizen Journalism: Locally Grown, Community-Owned*. Retrieved from http://www.stanford.edu/dept/undergrad/cgi-bin/drupal_pwr/sites/default/files/common/docs/BoothePWR1112Ho.pdf
Kokenge, J. (2010). *Why People Produce Citizen-Journalism: A Qualitative Analysis*. Master's thesis, University of Missouri. Retrieved from https://mospace.umsystem.edu/xmlui/bitstream/handle/10355/9289/research.pdf?sequence=3
Korgaonkar, P.K., & Wolin, L.D. (1999). A Multivariate Analysis of Web Usage. *Journal of Advertising Research*, 39, 53-68.
Jarvis, J. (2006, July 5). *Networked Journalism*. Message posted to http://www.buzzmachine.com/2006/07/05/networked-journalism/
Lasica, J.D. (2003, August 7). What is Participatory Journalism?. *Online Journalism Review*. Retrieved from http://www.ojr.org/ojr/workplace/1060217106.php
Lewis, C.S., Kaufhold, K., & Lasora, D.L. (2010). Thinking About Citizen Journalism:. The Philosophical and Practical Challenges of User-Generated Content for Community Newspapers. *Journalism Practice* 4(2), 163-179.
Li, D. (2005). *Why do you blog: A Uses-And-Gratifications Inquiry into Bloggers' Motivations*. Master's thesis, Marquette University. Retrieved from http://www.researchgate.net/publication/34665184_Why_do_you_blog__a_uses-and-gratifications_inquiry_into_bloggers'_motivations_
Lievrouw, L. (2011). *Alternative and Activist New Media*. Cambridge: Polity Press.
Littau, J. (2007). *Citizen Journalism and Community Building: Predictive Measures of Social Capital Generation*. Master's thesis, University of Missouri.
Mazzoleni, G., Vigevani, G. & Splendore, S. (2011). *Mapping Digital Media: Italy. A report by the Open Society Foundation*. Retrieved from http://www.opensocietyfoundations.org/sites/default/files/OSF-Media-Report-Italy11-07-2011-WEB.pdf

Metzgar, E., Kurpius, D. & Rowley, K. (2011). Defining Hyperlocal Media: Proposing a Framework for Discussion. *New Media Society,* 13(5), 772-787.

Nardi, B.A., Schiano, D.J., Gumbrecht, M. & Swartz, L. (2004). Why We Blog. *Communications of the ACM* 47(12), 41-46.

Padovani, C. (2010). Citizens Communication and the 2009 G8 Summit in L'Aquila, Italy. *International Journal of Communication* 4, 416-439.

Rucht, D. (2004). The Internet as a New Opportunity for Transnational Protest Groups. In M. Kousis & C. Tilly (Eds.), *Threats and Opportunities in Contentious Politics* (pp. 70-85). Boulder: Paradigm Publishers.

Schaffer, J. (2007). *Citizen Media: Fad Or the Future of News? The Rise and Prospects of Hyperlocal Journalism.* J-Lab: The Institute for Interactive Journalism. Retrieved from http://www.j-lab.org/_uploads/downloads/citizen_media-1.pdf

Schmidt, G. (2010). Web 2.0 for Disaster Response and Recovery. *Journal of Web Librarianship* 4(4), 413-426.

Sheehan, K.B. (2002). Of Surfing, Searching, and Newshounds: A Typology of Internet Users' Online Sessions. *Journal of Advertising Research*, 42(4), 62-71.

Taylor, M., Wells, G., Howell, G. & Raphael, B. (2012). The Role of Social Media as Psychological First Aid as a Support to Community Resilience Building: A Facebook Study from 'Cyclone Yasi Update. *The Australian Journal Of Emergency Management* 27(1), 20-26.

Treré, E., & Bazzarin, V. (2011). Exploring Italian Micro Web TVs: how high-Tech bricoleur redefine audiences?. *ESSACHESS Journal for Communication Studies*, 4(1), 49-67.

Treré, E., & Farinosi, M. (2012). *(H)earthquake TV:* "People Rebuilding Life After the Emergency". In A. Abruzzese, N. Barile, J. Gebhardt, J. Vincent & L. Fortunati (Eds.) *The New Television Ecosystem* (pp. 61-79). Berlin: Peter Lang.

Vettehen, P. H. H., Renksorf, K. & Wester, F. (1996). Media Use as Social Action: Methodological Issues. In K. Renckstorf, D. McQuail & N. Jakowski (Eds.), *Media Use as Social Action.* London: Libbey.

Weber, M. (1907/1956). *Wirtschaft und Gesellschaft: Grundriss der Verstehenden Socziologie*. Tubingen: Mohr.

Part III.

City Representations, Media Imageries and Urban Experience

CHAPTER XII

CINEMATIC ARCHITECTURES *INSITU*: NOTES ON THE PARTICIPATORY CONSTRUCTION OF A VISUAL URBAN IMAGERY

MIRIAM DE ROSA

1. Media, urban space and the imaginary: terms and prolegomena of the research

Similar practices in different contexts reveal common *motifs* and highlight macro-trends. In this sense, the thematization of the city operated by contemporary media and, vice versa, the permeation of urban territory by a full set of media expresses the deep relationship between media and the city. This is such an intertwined, global-scale connection that sometimes it becomes hard to distinguish where the influence of media on the urban element stops and where the inverse process begins. On the contrary, what is extremely clear is that they both are symptoms of a mutual macro-trend–on the one hand our urban imagery encompasses media, while on the other hand our medial imagery somehow includes the widespread diffusion of medial tools, *dispositifs* and practices throughout urban fabric.

In this context, the moving image plays a central role, as it has historically been a pioneering form of giving a faithful representation of the city[1], and it recently started to work as basic technology for a large

---

[1] From the first panoramas to the Lumière views (*La sortie de l'usine Lumière*, 1895; *L'arrivée d'un train en gare de La Ciotat*, 1896), from Dziga Vertov's *A Man with a Movie Camera* (*Čelovek s Kinoapparatom*, 1929) to the famous city symphonies (with a rather evident reference to Walter Ruttmann's *Berlin: Die Sinfonie der Großstadt*, 1927), the city has always been a privileged cinema set.

number of further experimentations which transform the city in a sort of contemporary movie theatre (McQuire, 2008; Casetti, 2009; De Rosa & Franchin, 2009; McQuire, Martin & Niederer, 2009; Koeck & Roberts, 2010).

This study assumes these preliminary considerations and presents a focus on a specific media production in order to explore in further depth the link relating media and the city, reserving a particular attention to the cinematic sphere.

Therefore, in an analytic perspective, the “media-city bond” develops at least on two different levels:

- On a production level, this connection works as the basis of a number of cross-media products, leading to a complex media platform and a multifaceted media format able to pervade cities.
- In terms of content, and thus on a representational level.

The interactive film *INSITU* presents both these aspects. Moreover, it gives the chance to observe how these two levels intersect, since it synthesises their convergence into a symbolic frame.

Examining the film, I want to test the following hypothesis: the formal and essential[2] layers defining the tight relationship between media and the city not only can find a meeting point, but also concur in shaping a coherent universe of sense. In particular, the territory where these two elements seem to interact is the dimension of the imaginary, that is to say an articulated *ensemble* fed by meanings, experiences and practices. In other terms, my aim is to demonstrate the importance of the imaginary, as the “horizon that determine[s] what we experience and how we interpret what we experience” (Crapanzano 2004)[3]. More specifically, I will try to show the quality of the selected case study as a visual database storing frames of a global, lived, mediated city, able to enhance and rearticulate the current urban imagery. In fact, thanks to a participatory platform, *INSITU* encourages to share one’s own experiences of the city and their cinematic rendering, enabling the construction of a sophisticated

---

[2] I use here *essential* in a philosophical sense, meaning something connected to the nature of the subject, and thus to its content.

[3] On the notion of imagery and its connections with the cultural and media industry, please refer to Brancato, 2000.

architecture, which weaves medial and urban fibres into a complex symbolic whole.

2. The project *INSITU*

INSITU is an interactive film by Antoine Viviani produced by Arte and Providence. Presented as a collective project, the film wants to be a "poetic documentary about the urban space in Europe nowadays"[4]. Pursuing this goal, it inevitably works as a storytelling form, for it offers a sort of *mise-en-scène* of the artistic and experimental interventions found throughout our cities.

Starting from the authorial idea by Aurélie Florent and the directorial figure of Viviani, the structure of the film includes chapters released by different users[5], a fact that makes it possible to define the production as a kind of "post-textual and open-source film". Despite the presence of a post-production work team, the philosophy which supports the project envisages a variable structure which can be continuously integrated. The result is an archive and at the same time an open text (Eco, 1962), strategically arranged according to a (partial) "bottom-up" logic (Bruns, 2008; De Blasio & Peverini, 2010) in order to preserve this very potentiality to be expanded, enlarged, enriched. This filmic grassroots production (Fanchi & Casetti, 2006; Abruzzese & Ferraresi, 2009; Fanchi, 2010; Wittke & Hanekop, 2011) represents just the cinematic part of a whole that encompasses a wide range of digital channels, including a website[6] and a blog[7], an interactive online map[8], a mobile application[9] and a presence in the social media environment[10].

---

[4] From the interview to Antoine Viviani, available at cinemadocumentaire.wordpress.com/2011/12/12/insitu-entretien-multimedia-avec-antoine-viviani/ (last accessed March 28th, 2012; my translation).

[5] This recalls the fruition model typical of the book, subsequently assumed by the DVD. Like these two media forms, the movie evokes the same structure because even in this case the spectator has a menu where he can select a specific section or a chapter. On the DVD and its features, please refer to Quaresima, L., & Re, V. (2010). *Play the movie. Il DVD e le nuove forme dell'esperienza audiovisiva*. Torino: Kaplan.

[6] insitu.arte.tv/en/#/home (last accessed March 26th, 2012).

[7] insitu.arte.tv/blog/?p=1074 (last accessed March 26th, 2012).

[8] insitu.arte.tv/en/#/map (last accessed March 26th, 2012).

[9] itunes.apple.com/us/app/arte-insitu/id455938432?mt=8# (last accessed March 26th, 2012).

[10] https://twitter.com/#!/insitufilm; www.facebook.com/INSITU.arte.tv (last accessed March 26th, 2012).

Such an articulated platform attests the cross-media intention of the project production. Under a technical profile, one's own videos, pictures and audio files can be easily shared thanks to a visual interface (Bijker & Law, 1992; Anceschi, 1993; Zinna, 2004) that allows uploading them simply by positioning them on the map.

The collective dimension of the film sources has an evident impact on the language and the style of the final product. In fact, the aesthetics of the whole work is a mix of documentary, socio-cultural investigation, filmic diary and video testimonies[11]. Nevertheless, each and every chapter succeeds on the one hand in providing rich materials devoted to media representation of cities and of those social processes taking place in the urban space and, on the other, in focussing on cities in a very clear way, highlighting the tight connection between them and media. However, many questions remain to be answered: how does *INSITU* sketch the outline of the city? How are media present throughout urban space, and in which way are they used to describe it? What kind of materials does the project permit to gather, and which kind of observations do they allow? What type of elaboration do the film and its mechanisms of production encourage?

3. Filmic explorations throughout the city

What *INSITU* basically grants both in terms of media format and content is an exploration. On the former level, the project builds a usage pattern made up of a number of prolifically integrated means, software and apps. In fact, taking into consideration the use of different devices and the development of the diverse practices which are implied, it gives a good chance to examine the possibilities offered by a cross-media experience. The logic inspiring the production is that of opening a platform, distributing the access to it across the contemporary media scenario. Thanks to the network linking the web site, the blog, the social media accounts, the mobile app and the online movie, the subject can benefit from such a media environment, which virtually adheres to his urban habitat[12]. The superimposition of these two structures gives birth to a space, better yet–a *mediaspace* (Couldry & McCarthy, 2004; Eckardt et

[11] Anyhow, the documentary is definitely the most proper genre to define the film, since it recently won the "Grand Pix du Digital Storytelling du Doc Lab" at the IDFA (International Documentary Film Festival Amsterdam).

[12] On the concept of *habitat* as "niche" practiced by humans, please refer to Rossi, 1985.

al., 2008) where to live. It is not a linear dimension, rather an articulated one, since the subject can move throughout it following his own path, therefore having a custom-made experience despite the unavoidable mediation. The opportunity to personalize the use of the platform is supported by the co-creative nature of *INSITU* and by the autonomy that such a dynamics confers to the movie and the map.

Involving a series of tools and devices which work synergically in order to converge towards an infrastructure that virtualizes a service (i.e. once again, the fruition of video contributions, the visualization of the map or the participation to the project by submitting what can then become an episode of the film), the project takes advantage of the possibilities given by the so-called cloud-computing[13] in terms of access and infrastructure scalability.

Also as far as the content is concerned, the representation of the city seems to be founded on a visual exploration. Indeed, the cinematic *dérive* experienced by the authors taking part to the project presents exactly their *flâneries*, their observation of the changes affecting the urban scenario, their interventions on the landscape and the neighbourhood, their engagement towards the space they live *in*.

The perspective adopted in this double exploration is thus a phenomenological one, since the only condition to be respected by the subject is actually to be *in situ*. Such an element emphasizes the geo-location of the individual and associates it with the ability to render the perception and the subsequent elaboration of urban space. Therefore, what the project seems to promote is a specific view of the city, which should necessarily be experienced "on site/on sight"[14]. The main mechanism is

[13] This is not the place to deal with cloud computing; on this issue, please refer to Buyya, R., Broberg, J., & Goscinski, A. (2011). *Cloud Computing: Principles and Paradigms*. New Jersey: Wiley&Sons; Antonopoulos, N., & Gillam, N. (2010). *Cloud Computing: Principles, Systems and Applications*. London: Springer; Gilje Jaatun, M., Zhao, G., & Rong, C. (2009). *Cloud Computing: First International Conference. CloudCom 2009. Beijing December 2009, Proceedings*. Berlin, Heidelberg: Springer.

[14] The terms *site* and *sight* have been coupled in a brilliant study by Evgenia Giannouri, *Site - Non Sight: Les paysages masqués de Blow Up* (paper presented on June 29th, 2009, at the "Université d'Été 2009" held at INHA, Paris). I adopt her pun here, trying to go deeper and to analyse the correspondence between a position *on site* and the chance to get a *sight* from it.

primarily the combination of the location of the subject in space, the relocation (Casetti, 2008) of (cinematic) media and the possibilities to produce and consume media contents in mobility (Brown, 2001; Green, 2002; Scifo, 2006). In this sense, the walk becomes the central creative action. Re-proposing a brand new, technologised version of an old practice, *INSITU* places itself along the vector starting from the Dada and Surrealistic Movement and passing through meaningful experiences like the formal constitution of the Lettrist International and Situationism, or the later development of Land Art[15]. As in these artistic traditions, the project conceives walking as a repository of a deep aesthetic value (Davila, 2002; Careri, 2006): it is a modality to feed–but also analyse and reflect upon– the relationship of the subject towards the environment, a way to understand and represent it.

It is precisely by means of walking that the interactive movie sketches the outline of the city. In fact, the different chapters of the film are released during personal explorations of the urban space, following someone wandering around or trying to thematise the idea of moving across the city. The authors are thus complex figures expressing thoughts and impressions. On the one hand they embody Michel de Certeau's stroller (1990), inventing their own everyday in catching it through the filmic device, while on the other hand they inherit the interest and the approach of the *flâneur*. Their paths throughout the urban space then become their way to possess the environment, impressing the trace of their presence inside it. A strong sense of marking the territory enlivens and stimulates such an action: entering a certain urban space, the subject makes it the place[16] of his experience of the city. Moreover, this conception clearly underlines a parallel between this "writing on space" given by the journey, and the releasing of the movie intended as "cinematic writing". Even under a productive light, a creative element seems to associate the two writings, because the territorial aspect and the filmic aspect end up superimposing, giving birth to a representational rendering of the city.

The materials collected by *INSITU* are the output of a subjective immersion in the urban space, a contemporary, personal, visual immersion. The result is very far from those products characterized by a

---

[15] A brief systematization of these experiences is presented in Careri, 2006.

[16] I use here the terms *space* and *place* as Martin Heidegger puts them (for a brief definition and an application of the concepts to the filmic device, see De Rosa, 2008; this interpretation is diametrically opposed to de Certeau's.)

cinematographic aesthetics based on excitement and wonder, on an almost Futuristic speed, on the taste for the unexpected and shocking (as in the city symphonies of last century). The style of the episodes composing the movie and the editing of some of them do not create an image of the urban chaos, rather the eye of the camera gives back the life and the rhythms of the city alternating them with a more intimistic observation of spaces, panoramas and citizens. All the images gathered by Viviani resemble a gallery of urban places, meaning just city portions turned into subjects' habitats by their walking across them, their capturing them through sight and sensible perception or any activity enabling to affix a personal connotation to space. This is an involvement in the urban territory that does not come from an image of the city as a hypnotic whirl[17], on the contrary what seems to link the urban environment and the authors is an immersive engagement, as the public and the personal sphere resonate in a mirroring game, where the cinematic image works as filter. Such a sensation is based on the feeling of being part of a space in which the position of the body, the experience, the visual and haptic perception (Sobchack, 1992, 2004) draw a particular image of the city. This is a visual construction that is obviously supported by the eye, but whose premises are to be found in the complex dimension of the full perception (Malavasi, 2009)–here is the reason why I find the on site/on sight correspondence particularly emblematic. Indeed, this association exactly emphasizes a modification of the urban imaginary as it is represented by cinema. The stress on the mobility of the gaze and of the camera perspective, the imperative, almost mandatory research of dynamism thanks to a particular frame composition and fast editing are finally abandoned, in favour of a more anthropological look: the lady scratching the Nazi stickers off poles and street signs (ch. *Irmela goes to war*; figures 1-2) denotes a singular but strong intervention on the city and a first person action on the urban surface which becomes something to care about.

[17] Such was the cinematic representation of the city at the beginning of the 20th century.

Figure 1 - Stills from the film chapter *Irmela goes to war* © *Providences*.

Figure 2 - Stills from the film chapter *Irmela goes to war* © *Providences*.

Figure 3 - Stills from the film chapter *The echoes of Alain Damasio* © *Providences*.

The pseudo-philosophic reflection about the anthropization of urban space (ch. *The echoes of Alain Damasio*; figure 3) indicates an acute sensitiveness and a high awareness of the importance of men's role in designing, building, running the city. The excitement of mazes and roller coasters typical of a surprised and easily impressionable approach to the city is replaced by a sort of visionary utopia evoking just the opposite: "it's really all about inventing a slow leisure activity", we hear in a chapter of the film (*Nogo voyages*; figures 4-5-6), as if the inhabitant imagined by the voiceover would rather spend his time in an "augmented-circulation"[18] society able to preserve some space to meditate and live calmly, than being fascinated by a city thriving on "the hustle and bustle of a crowd in the traffic", no matter if it identifies "a metaphor of the decay of contemporary culture"[19] or maybe represents something to idealize and romanticise, as in Allen's *Manhattan* (1979).

[18] The last quotations are taken from *Nogo voyages*; my translation.
[19] These and other attitudes towards the city characterised the perception of the urban environment throughout modernity and in the whole of the 20th century. In particular, I'm echoing here the famous opening sequence of *Manhattan* (Woody Allen, 1979).

Figure 5 - Still from the film chapter *Nogo voyages* © *Providences*.

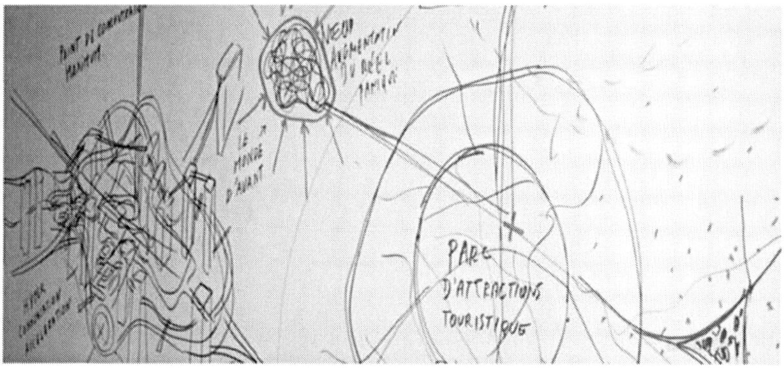

Figure 6 - Still from the film chapter *Nogo voyages* © *Providences*.

The film thus focuses on some questions affecting the contemporary city, as the renovation of interstitial spaces, the requalification of outer areas, the fluxes of people and their living spaces, the functionality and fluidity of the connection solutions among faraway neighbourhoods, giving birth to a thick layer of images superimposing one another, able to renovate and recreate the idea of the city. Hence, the visual representation of urban space and the filmic testimonies collected by *INSITU* not only work as chronicle of the evolution of the city, but also trace the modification of our urban imagery. The kind of observation that these cinematic materials enable is a grassroots ethnography of the city, that tries to take into account the ways in which urban spaces are turned into places, their uses and the sociocultural practices ruling the public and personal spheres (Casetti, 2005; Marrone & Pezzini, 2006).

This brings us to the heart of the reflection: as the project shows, the visual representation of the city directly influences the structure of our urban imagery. The tight connection linking the filmic and the symbol elements resides in the imaginary. The moving image helps therefore in visualising what generally remains in this abstract dimension, offering a representation of "what enables, through making a sense of, the practices of a society […] the ways in which people imagine their social existence, how they fit together with others" (Taylor, 2002) and depict their urban context. In this perspective, the authors participating to the interactive film embody something more than their *flâneur* ancestors:

> These individuals constantly increase their sociological relevance, as they become emblematic figures of a late-modern society affected by deep transformations. Those changes impact particularly on the individualization processes of human experience, the composite articulation of identification trajectories with the territory, the diffusion of daily reflexivity practices[20].

An emphasized elaboration and re-elaboration of personal experience, a new coordination between individual and collective action, the possibility to share these reflections make the contributors to Viviani's project a more sophisticated version of the Baudelairian wanderer. A clear dimension of ordering and reordering reality, the experience of the city and the world, and a more delicate and sometimes fragile perception of the Self characterize the authors. They add to their view a deeper awareness of the image, of its potentialities and its strategies for use, as if they could profit from their own habit and education to code, represent and interpret the swarming visual scenario featuring their everyday life.

Among other reasons, such a revisitation of the *flâneur* figure is made possible by the rich set of media choices available to the contemporary city-walker. In this sense, the cross-media platform produced for *INSITU* is an example, and the interactive form of the film is another possibility to express the values, the emotions and the meanings of one's own strolling as well.

The moving image enters therefore a complex symbolic construction, contributing in the creation of a new audiovisual geography of the city,

[20] Nuvolati, 2006.

which reverberates a further articulated symbolic structure–the architecture of the imagery.

4. Conclusions: Cinematic Architectures and visual urban imagery

The movie and its mechanisms of production encourage a particular elaboration of the materials collected in the filmic archive supporting the project. It seems to be a creative release based on a sharper confidence with introversion and extroversion processes. The results are detailed descriptions of the urban fabric, where the subject's biography filters the impression of the space, its image, its dimensions and features (Fanchi, 2005); the filmic observation of the events happening in the city becomes something useful in defining not only the context but also the identity of the ones taking part or somehow connected to them; urban sights and the whole range of perceptions are processed and reprocessed afterwards, shaping the image of what they do illustrate and articulating a space which is mediated and re-presented through the image. Therefore, the media-city rises as the horizon of the contemporary Self in space, where social knowledge and medial forms contribute to the creation of new content configurations (Fanchi & Villa, 2011). In this sense, the *mediaspace* is to be seen as extroversion of the imagery, and therefore close to the co-construction of the urban imagery itself. In the same context, the filmic device promotes important recollections about the city, considering diachronically both the historical evolution and the most recent images of the urban situation. This is basically the starting point of a collective visual imagery, which encompasses new representative modalities and the most traditional, rhetoric cinematographic narrations.

Moreover, the experimentation granted by the innovative character of *INSITU* competes in a certain way with the preservation of the urban heritage as the citizens conceive it. The appearance of particular places of the city (i.e. ch. *The white zones of Philippe Vasset* referring to Paris, and ch. *Explore the heritage – Berlin*) or the cinematic rendering of specific urban aspects (i.e. the soundscape of ch. *The city resonates*) focus on peculiar aspects belonging to and feeding this imagery: "the vision of natural and urban places, real or dreamlike, is elaborated through those sophisticated mechanisms able to turn it into an image" (Bordini, 2010, p. 26). The time of the city, its rhythm, its acoustic profile or its curiosities are similarly involved in a mechanism of elaboration, able to visualize these aspects, manipulating them according to individual or collective perceptions, sensory involvement or cultural formulations. Such a resultant construction leads to a multifaceted ideal architecture.

Cinema has been traditionally the vehicle of the images converging in this construction, and at the same time, it has contributed to stimulating, changing, updating it. *INSITU* assumes this same ability, trying to work as a kind of "cinesocial" device able to work as cultural interface (Manovich, 2008).

To sum up, the case study shows how the mediated connotation of the spatial dimension establishes new public and private spheres, and defines the position of man in space. In this sense, the media regulate the approach towards the urban element, for they work as vehicle of introversion/extroversion of the Self, encouraging a personal mapping of space.

INSITU represents a privileged lens to look from the inside at the transformations reshaping urban environments, the contemporary mediascape, and the imagery of both. The project shows thus how the image of urban space is something articulated and constructed through the elaboration and re-elaboration, the representation and symbolization of the city. Urban imagery is indeed a visual architecture built increasingly often by means of media usage, whose "genealogy" (Carmagnola & Matera, 2008) is to be traced back to the mediated experience of cities.

References

Abruzzese, A., & Ferraresi, M. (2009). *Next. Identità fra consumo e comunicazione*. Bologna: Lupetti.

Anceschi, G. (1993). *Il progetto delle interfacce*. Milano: Domus Academy.

Bijker, W., & Law, J. (1992). *Shaping Technology/Building Society*. Cambridge, MA: MIT Press.

Bordini, S. (2010). *Appunti sul paesaggio nell'arte mediale*. Milano: Postmedia Books.

Brancato, S. (2000). *Sociologie dell'immaginario*. Roma: Carocci.

Brown, B. (2001). *Wireless World: Social, Cultural and Interactional Issues in Mobile Communications and Computing*. London: Springer.

Bruns, A. (2008). *Blogs, Wikipedia, Second Life, and Beyond: From Production to Produsage*. New York: Peter Lang.

Careri, F. (2006). *Walkscapes. Camminare come pratica estetica*. Torino: Einaudi.

Carmagnola, F., & Matera, V. (2008). *Genealogie dell'immaginario*. Torino: UTET.

Casetti, F. (2008). L'esperienza filmica e la ri-locazione del cinema. *Fata Morgana, Esperienza, 4*, 23-40.

—. (2009). Filmic experience. *Screen, 50* (1), 56-66.

—. (2005). *L'occhio del Novecento. Cinema, Esperienza, modernità*. Milano: Bompiani.

Couldry, N., & McCarthy, A. (2004). *Mediaspace. Place, Scale and Culture in a Media Age*. London, New York: Routledge.

Crapanzano, V. (2004). *Imaginative horizons: an essay in literary-philosophical anthropology*. Chicago: University of Chicago Press.

Davila, T. (2002). *Marcher, créer. Déplacements, flâneries, dérives dans l'art de la fin du XXème siècle*. Paris: Regard.

de Certeau, M. (1990). *L'invention du quotidien. I. Arts de faire*. Paris: Gallimard.

De Blasio, E., & Peverini, P. (2010). *Open cinema. Scenari di visione cinematografica negli anni '10*. Roma: Ente dello Spettacolo.

De Rosa, M. (2008). To look, to wander: Cinema in Installations. *Cinéma & Cie. Relocation*, Casetti, F. (Ed.), 11, Carocci, Roma, 31-39.

De Rosa, M., & Franchin, G. (2009). Forme dell'abitare. Pratiche di tracciabilità tra mondo e reale. *Comunicazioni Sociali on line, 1*, available at http://www.comunicazionisocialionline.it/2009/1/4/loadPDF/.

Eckardt, F., Geelhaar, J., Colini, L., Willis, K. S., Chorianopoulos, K., & Henning, R. (2008). *Mediacity – Situations, Practices and Encounters*. Berlin: Frank&Timme.

Eco, U. (1962). *Opera aperta: forma e indeterminazione nelle poetiche contemporanee*. Milano: Bompiani.

Fanchi, M. (2005). Immaginari cinematografici e pratiche sociali della memoria. *Comunicazioni Sociali, 3*, 489-496.

—. (2010). Iperboli, crasi e neologismi. Per un'onomastica (e una teoria) del "nuovo" spettatore. *Cinergie. Il cinema e le altre arti, 19*, 58-59.

Fanchi, M., & Casetti, F. (2006). *Terre incognite: lo spettatore e le nuove forme di esperienza di visione del film*. Roma: Carocci.

Green, N. (2002). On the Move: Technology, Mobility, and the Mediation of Social Time and Space. *The Society, 18*, 281-292.

Koeck, R., & Roberts, L. (2010). *The City and the Moving Image: Urban Projections*. Basingstoke: Palgrave Macmillan.

Manovich, L. (2008). *Cinema as a Cultural Interface*. Text available at http://www.manovich.net/TEXT/cinema-cultural.html (last accessed, March 29th, 2012).

McQuire, S., Martin, M., & Niederer, S. (2009). *Urbanscreens: Reader*. Amsterdam: Institute of Network Culture.

McQuire, S. (2008). *Media, Architecture and Urban Space*. Los Angeles, London, New Delhi, Singapore, Washington DC: Sage.

Malavasi, L. (2009). *Racconti di corpi. Cinema, film, spettatori*. Torino: Kaplan.

Marrone, G., & Pezzini, I. (2006). *Senso e metropoli. Per una semiotica posturbana*. Roma: Meltemi.

Nuvolati, G. (2006). *Lo sguardo vagabondo. Il flâneur e la città da Baudelaire ai postmoderni*. Bologna: Il Mulino.

Rossi, L. (1985). *La nicchia ecologica: teoria ed applicazioni*. Roma: Istituto della enciclopedia italiana.

Scifo, B. (2006). *Culture mobili. Ricerche sull'adozione giovanile della telefonia cellulare*. Milano: Vita e Pensiero.

Sobchack, V. (1992). *The Address of the Eye: a Phenomenology of Film Experience*. Princeton: Princeton University Press.

—. (2004). *Carnal Thoughts. Embodiment and Moving Image Culture*. Berkeley, Los Angeles, London: University of California Press.

Taylor, C. (2002). Modern Social Imaginaries. *Public Culture*, *14* (1), 281-292.

Wittke, V., & Hanekop, H. (2011). *New Forms of Collaborative Innovation and Production on the Internet An Interdisciplinary Perspective*. Göttingen: Universitätsverlag.

Zinna, A. (2004). L'objet et ses interfaces. *E|C Rivista on line dell'Associazione Semiotica Italiana*. Available at http://www.ec-aiss.it/archivio/tipologico/autore.php (last accessed March 29th, 2012).

CHAPTER XIII

DIGITAL URBAN IDENTITIES

KATALIN FEHÉR

Introduction

Each city is in possession of a representative and meaningful digital identity. This identity is complex and interactive: it implies and entails patterns of human urban participation, "the experience of a city". It also implies and entails the interpersonal profiles required for communication in urban scenarios. This complexity of representation is constituted by meaningful data complexes processed online via the metropolis of digit(al) convergence and divergence and operates the "common" platforms of the city online.

Urban digital representation constitutes a (or the) digital identity of a town and is organised via multiple networking. Online city networks are trafficking "city contents", including real time, real space people who are online and/or are represented within these networks–some of them are communicated via professional (marketing) communication and user-generated contents. The offline reality of a city is being substituted by its online representations. Flesh and blood, stone and concrete become an image/icon online. This networking environment is a *virtual blazon* of the city, its space and timing are interfaces within networking: it defines symbolic loci and events (time) with platforms and applications and most probably focuses on social-cultural collaboration real time in and among public places. These are the common places of the urban algorithms. This virtual Other–that is the online "version" of the offline f/actual real city–may provide some reflection for offline real time life (e.g. geo-tagging), and, also, for various virtual projections of the city (c.f. "Second Life").

Cities represent, communicate and mediate their operations in real time both offline and online. The city is the media and this defines its attributes, organises its loci.

This paper interprets the city within online digital new media environments as a digital identity with branded contents for collaboration. Users are networking in multifarious ways–and the online chains of channels along which each and every user can "netwalk" provide them with chances of engagement.

1. The city online I. Identities in the making

The city has users. Cities' identities belong to their users or are generated by their users' identification. Who was the most famous/greatest inhabitant of this place? Who is living there now? Who is checking in now via geolocation devices? Who is the commuter and who is the visitor? Who is the inhabitant and who is the tourist?

A city does not exist without references to human identities. If somebody, or the city, refers to somebody else it may generate human traffic, and it may also generate the lack of human traffic. Our identification is also related to others and to others' locations. This is one reason for not being there.

In real time, offline, we can be only in one place at the one time, but we may have multiple parallel mental and virtual locations. The starting point is our own identity, which is the main point of reference: it is the subject that frames the self for networking. We set up connection from this zero point with other links: visible and invisible people in different places. Visible and existing people are in real-time connections offline and online, however, there have also been invisible iconic subjects (that is, icons) in local history who guide us, they have been the (earlier) constructors of loci, they built the stage on/offline.

Human identities anchor links in a city. They generate and manage their places, loci and changeable narratives/references in the social-cultural context. They provide the images and rhythms of a city. Their loneliness in the car, their common ways when participating in public platforms/places, their consumption of news concerning politics, economy or sports, local marketing goals and their relatedness to social-cultural opinion leaders build up the identity of the city via social and communicative interactions of subjects.

The strategy is mediatised. Human identities use urban areas for everyday practices and for fun. Random/real time social networks are expectant of sharing. Flashmobs, civil or marketing actions, platform-dependent signs/symbols redefine experience in the city.

Human identities can be organisations, groups, people. Professional and user-generated contents and intentions blend in a socio-cultural collaboration. Technology and marketing define public spaces and privacy via identified usability. City places are experienced loci. There are *refugees* (Sennett, 1994) who are provided with identities via social loneliness. The goals are vitality, readiness to visit, to entertain places for urban human identities so as to link and to crowdsource. If it is "cool and sexy" a city's identity attracts human identities and vice versa. They would like to belong to this temporarily or permanently, to be a real or a virtual part of it. A good example is that of the "virtual mayor" Foursquare application: activity, consumption or regular stay in one place provide a role of regency/procuratorship. The user becomes a "mayor" of a place with a crown for some time, or longer, and virtually.

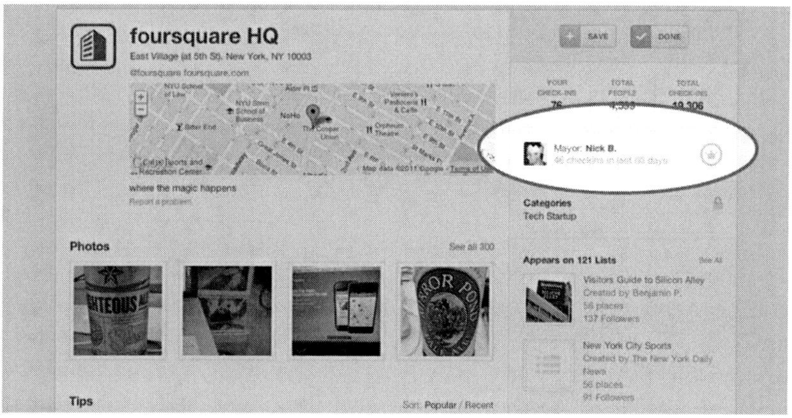

Figure 1 - Virtual mayorship on Foursquare. Source: Foursquare website.

If somebody owns a place, the place also owns him/her–via symbols, connections, log in, fingerprints/retina scanning and so on. This phenomenon refers to reciprocity among identities that pattern cities online. Cities become loci for refugees to get access to an urban identity when their users define them–in contrast to other cities.

2. The City Online II. Loci and events

A city has symbolic loci and events in public or (partly) closed real places/real time and parallel ones "on air" with live shows and tags. Geotagging and virtual projection relate visual elements and individual/common experience to archiving and sharing. The main goals can be navigation, featuring a point of view or an occasion, self(re)definition, finding practical or fun links for users and city guides, (public) service and transport, local marketing to communicate and organise a city. Loci and events are appropriated by organisational and human identities, public practice via experience, and adventure with online casting.

Harvey (1989) stressed *shrivelling* in global life. Events and objects are experienced as simultaneously imported from different space and time continuums. This shrivelling reflects McLuhan's term, "global village" (1964) where instantaneous movements occur via electric technology at the same time to create discontinuities, divisions and diversity. Fragmentation in a network society (van Dijk, 2007) means updated logistics to navigate on different loci and events. Fragments generate set ups and promote restructuring.

Sassen (2000) defines metaphorically the postmodern spatial structures as *thickening*–this refers to malls, hypermarkets and other similar phenomena. Users' spot sensibility focalizes these common and semi-public places so as to create thickening links. Construction will be real and virtual together without fixed meanings. Activities, events, symbols of places are settled according to updated agreements in the digital era, after postmodernism. Local traditions and fragmented (sub)cultures, blended forms of communication are in interaction and they are digitalised. Regional decisions and acts generate some final meanings: they may anchor the virtual in a local domain, reducing the number of choices the users who visit these postmodern spatial structures might have on account of their thickening media.

Manuel Castells rethought the earlier concept of *place of places* into the expression *space of flows*. "The space of flows [...] links up distant locales around shared functions and meanings on the basis of electronic circuits and fast transportation corridors, while isolating and subduing the logic of experience embodied in the space of places". (Castells, 2001, p. 155.) According to his theory, digital age implies dynamic interactions where flows are above historically constructed spaces via networks.

"Can You See Me Now?" is an electronic art project that is visible simultaneously both online and in the streets. The work was premiered in Sheffield at the b.tv festival. The conceptual background is provided by the means of mobile and 3G technologies and it has dramatic repercussions for the cities and for hybrid city landscapes. The overlapping real and virtual domains that are interlinked hypothesize a flexible logic for places and loci. Players represented spaces of flow online and offline. 20 people were playing online at the same time with different tactics and sending messages about moving to a control centre. Computers were showing the positions of offline players to guide them via audio streaming and to send them messages. Players registered instructions from virtual spaces of the city and made decisions with reference to real live situations. The flexible relationship to the real city mixed real/virtual tactics and generated overlaps. The common cognitive knowledge concerning typical urban locations was the most important constituent of it.

Most of the events and loci become similar. In a global context these meet and are selected in a global city's evolution. Global villages imply convergence in geo-design, urban design and other patterns of urban identities. Similar or identical public places (e.g. hypermarkets), products and services (e.g. McDonalds) exist. Globalisation results in homogenisation.

Urban identity needs to rethink and reinvent the genius loci of the place. Urban marketing and geoidentity, the sense of events and places, user-generated content are building via fragments in digital shrivelling and thickening, however they are looking for their identity in a possible divergence. Postmodernity merged all discourses (Jameson, 1991), and the new media have digitalised them. Thematic and spontaneous, traditional and retro, popular and niche are linked by users' identities (they are looking for their identities real time) and via their platforms and algorithms.

Figure 2 - "Can You See Me Now?" project, Dublin 2007. Source: humanitieslab.stanford.edu/44/278

3. The City Online III. Platforms and algorithms

Online-linked human identities fill up a city with life in this digital age. They use platforms to connect with loci and events, share their experience via social networks, follow events, visit geo-tagged locations, and define loci via acts or a lack of acts. Search engines, social media and other platforms give chance to generate urban identities. Collaboration, self-representation, visibility or invisibility establish interactions between the online and the offline in order to constitute a/the city. This implies a virtual or real space for data, and an electric space in interaction with extensions and with a structure of its own (Carazo-Chandler, 1998).

Objective, observed and mental space used to be associated with an alternative form of experience: we were aware of its virtual nature and we took it for granted that this objectively observed mental space denoted actual location based services offline. Spaces of flows are generated, finally, also, by users and algorithms. Algorithms build visible and mainstream blazons for cities. The first three Google hits turn the traffic for landing pages and they refer to a real time image of a city (reflecting history, tourism, economical risk, etc.). A typical navigation system gives three transport options (car, public transport, walking). Event organiser services also provide three choices to choose from (yes, no, maybe). Users vote by making their choices with reference to places, events and others digitally (like, unlike). Two or three choices organise patterns of platforms, flows of data and messages, decisions, thickenings, convergence and entropy.

Augmentation mediates between real and virtual spaces. Augmented reality provides a frame for activity and consumption in a city or for a city lifestyle. The frame can be a device or a platform with algorithms. A virtual supermarket in a subway station is a case in point where users shop by scanning QR codes on their smartphones: this is the case of augmentation. This means a virtual store that operates via displays, the real waiting time that is saved by virtual shopping may come handy for public transport. The service is a “homeplus” service provided for metropolitan lifestyle by Tesco. This marketing and sales technique has been increasing their sales.

Figure 3 - Virtual supermarket in a subway station, South Korea 2011. Source: www.littledoremi.com/tesco-homeplus-subway-virtual-store-in-south-korea/

The next layer of this can be Second Life (or some other platform of simulation) with screened copies to route marketing, and it can also be used in a parallel life. This projection also works for the past (New Orleans reconstruction before Katrina hurricane, 2005) or for fantasy (*Avatar*, 2009, is a science fiction movie featuring simultaneous loci: the film's celluloid universe is constituted by the mix of biosphere and simulation).

4. Consequences

What do we refer to with the phrase, the "virtual blazon" of a city? This is a feature where urban users and algorithms are linked via platforms and collaboration–including its negative forms, that is, conflicts. Virtual and augmented visibility, mobility and rhythms match the city with their patterns of identification. Professional and user-generated content and networks, flexible and temporary structures are linked in real time. Traditional, well-known and intentionally marked scenes come to life in interactions via online and virtual platforms. Users have plenty of choices to connect them to other platforms, rescue refugees in space of flows.

Cities are interactive media operated by user-activated urban identities and they are operated via digitally linked thickening. Digital communication reorganises and restructures urban space, the earlier (geographical, historical, social, cultural and economic) layers are going to be deconstructed. This is the making of *co/geography*.

References

Carazo-Chandler, C. (1998). *Cyberspace, Another Geography. Territories, Boundaries and Space*. New Zealand: University of Canterbury.

Castells, M. (2001). *Informationalism and the Network Society*. In P. Himanen (Ed.), *The Hacker Ethic and the Spirit of the Information Age* (pp. 155–178). New York: Random House.

Dijk, van J. (2007). *The Network Society*. Los Angeles: Sage.

Harvey, D. (1989). *The Condition of Post-Modernity*. Cambridge, Oxford: Blackwell.

Jameson, F. (1991). *Postmodernism or, Cultural Logic of Late Capitalism.* Durham, NC: Duke University Press.

McLuhan, M. (1964). *Understanding Media: The Extensions of Man*. New York: MIT Press.

Sassen, S. (1996). *Losing control? Sovereignty in An Age of Globalization.* New York: Columbia University Press.

Sennett, R. (1994). *Flesh and Stone. The Body and the City in Western Civilization.* New York: Norton.

Winston, B. (1998). *Media Technology and Society. A History: From the Telegraph to the Internet.* London: Routledge.

CHAPTER XIV

THERE IS A LIGHT THAT NEVER GOES OUT. CULTIVATION EFFECTS ON THE FREEDOM OF MOVEMENT IN URBAN AND SUBURBAN PUBLIC PLACES

GABRIELLA SANDSTIG

1. What are the consequences of Mean World Syndrome on behaviour?

Cultivation theory (Gerbner, 1972, 1973; Romer, Jamieson & Aday, 2003) focuses on the relationship between how heavy media consumption can, in the long run, alter our perceptions of reality to become stereotyped, distorted and selective. This perception can lead to our overestimation of the likelihood of becoming a victim of crimes or to a tendency to view the world as a mean or scary place (Mean World Syndrome) (Gerbner, 1980; Gerbner, Gross, Morgan & Signorielli, 1980).

Contemporary research has continued along this line (Cheung & Chan, 1996; Nabi & Riddle, 2008), but has also extended cultivation analysis to include a wide range of content in the areas of health (Chul-Joo & Niederdeppe, 2011), science (Bauer, 2005; Dudo, Brossard. Shanahan, Scheufele, Morgan & Signorielli, 2011), environment (Good, 2009), materialism (Shrum, Lee, Burroughs & Rindfleisch, 2011), state economy (Hetsroni, 2010) and genres like television talk shows (Woo & Dominick, 2001) or the focus has been placed on the methodological aspects of cultivation theory in the use of different heuristics or methods for sampling (Shrum, 1993, 1996, 1997, 2001a, 2001b, 2007; Buselle, 2001; Bradley, 2007).

However, even if contemporary research features studies on the relationship between violent media use or content and perceptions of reality as a mean or anxiety-causing place, much less attention has been paid to the possible consequences of these perceptions on behaviour, like limitations to the freedom of movement in one's city or neighbourhood. This is more common within the field of criminology. To measure levels of fear, criminologists have often built their analysis on the assumption that people's sensations of fear are based on a fear of crime (Hale, 1996). They also frame their analysis around the extreme case of asking respondents to what extent they avoid going out after dark because of fear, worry or insecurity. This type of framing does work, since analyses are optimized to provide maximum variation in the results. However the explanatory value of such a framing is low for levels of fear and insecurity in the everyday lives of people. First of all there are numerous studies that point toward the paradox of fear: that the majority of the people having experienced fear or insecurity themselves lack experiences of crime (Hale, 1996; Heber, 2005). Secondly, fear and insecurity in public spaces not only occur after dark but also during daytime, and the indicators of this type of fear and its mechanisms differ from the fear and insecurity sensed after dark, as does the sensation depending on the social as well as physical context of urban space (Sandstig, 2010a, 2010b, 2013). Thirdly, people try to avoid places they fear in their everyday lives, and if a municipality or neighbourhood aims to lower people's sensations of fear and insecurity, then other measures need to be sought to grasp the magnitude of potential attendance.

From a methodological standpoint this paper has two interests: nuancing the operationalisation of the limitations in the freedom of movement in public places; and comparing its empirical results with those of a previous study performed in the region of Västra Götaland and the municipality of Gothenburg as a whole (Western-SOM) (Sandstig, 2005, 2010a, 2010b). This since the population of respondents in this paper's study covers the population standing for the greatest statistical loss in the previous study (respondents with lower education, the socio-economically weak and those from a non-Swedish ethnic background), and therefore can contribute with additional knowledge.

The study of freedom of movement in urban public spaces and its limitations is also built upon the normative idea of equality, that all citizens are entitled to equal access to the city's public spaces. Feminist geographers like Doreen Massey (1994) have however shown that city spaces are marked by gender differences, and geographers like David Harvey (2001; 2009) and Don Mitchell (2004) that access to public spaces are restricted by socio-economic inequality.

It also warrants specific interest to compare the cultivation effects in a neighbourhood at the suburban level with the city or urban level as a whole, mainly because we know that media and news consumption is much lower in the socio-economically more restrained areas in focus for this study than in the city in general. The importance of cultivations effects due to media use are from this standpoint, to the extent that they occur and are comparable, expected to be lower than for the city in general. We also know from a previous study (Sandstig, 2010b) that the Mean World Syndrome, in the sense of perceiving the world out there as a scary place, does occur at the urban level. Mean World Syndrome originally refers to the research performed by Georg Gerbner and his colleagues working on indicators of cultural cultivation of heavy television viewing. Their research found that "[…] heavy viewers (compared to light viewers in the same social groups) derive from their television experience a heightened sense of danger, insecurity, and mistrust, or what we call the 'mean world' syndrome." (Gerbner, 1980, pp. 69-70). In the sense of Gerbner the "mean world" syndrome is the main result of mediated cultivation, where "[…] constant daily cultivation seems to add to a morbid sense of normalcy. Yet it is all well (if unwittingly) calculated to cultivate a sense of insecurity, anxiety, fear of the 'mean world' out there, and dependence on some strong protector" (1980, p. 66). However Gerbner, Larry Gross, Michel Morgan and Nancy Signorelli separated the sense of insecurity and fear from mistrust when elaborating on the operationalisation of, on the one hand, an Index of Perceptions of Danger, and on the other a Mean World Index (Gerbner et. al., 1980, p. 23). Here the latter is composed of three items that "measure the degree to which respondents agree that most people are just looking out for themselves, that you can't be too careful in dealing with people, and that most people would take advantage of you if they got a chance" (1980, p. 17). In the analysis the authors show that there is a strong correlation between the amount of television viewing and the tendency to express mistrust (1980, p. 17). In a meta-analysis of two decades of cultivation analysis, Michael Morgan and James Shanahan (1997) conclude that there is support for that the mindset of Mean World Syndrome among heavy viewers correlates with the same groups' general mistrust of people (Morgan & Shanahan, 1997, pp. 1-46). Both the perceptions and sensed experiences of insecurity and fear (similar to Gerbner et. al. perceptions of danger) as well as mistrust towards other people in general (in the sense of Morris Rosenberg's faith in people but in the use as the 11-point scale introduced by the SOM-Institute in 1995) are plausible and possible indicators to analyse in correlation to the limitations of the freedom of movement in focus of this paper. In this paper, and the study previously referred to as compared to, yet another indicator for the "mean world" out there is used. The indicator used is the perception of crime coverage of the media as being underrated, fair or overrated. Due to the lesser use of media in the suburban areas under study in this paper, the occurrence of Mean World Syndrome is also expected to be lower at the suburban level.

We also know from the same study that the sensation of experiencing fear and insecurity is lower at the suburban level than at the urban level, and that crime rates are higher in the city centre than in the suburbs of the town of Gothenburg. We know that both cultivation, resonance and enhancement effects due to personal, social and media related experiences of threats and risk as well as Mean World Syndrome has cultivation effects on the limitations in the freedom of movement at the urban level, while it is expected that this also has an effect on the suburban. What we don't know is if these effects are stronger, the same or weaker than at the urban level.

It has not further been analysed whether cultivation, resonance and other enhancement effects occur on the limitation on the freedom of movement at the suburban level of public spaces, in everyday life. This paper analyses whether cultivation, resonance and enhancement effects on limiting the freedom of movement also occur at the suburban level, in those areas of the town were the consumption of news media is among the lowest, and if so, how. The research investigates the similarities and differences:

- in the limitations of freedom of movement in the suburban versus urban spaces;
- in personal, social and mediated experiences of violence and crime;
- in cultivation, resonance and enhancement effects due to mediated experiences of crime and violence;
- in cultivation effects due to the perceptions of crime coverage in the media.

2. Cultivation and resonance theory and enhancement effects

2.1 Cultivation theory

Cultivation theory has taken an interest in the socialization effects of the media, how we learn the norms of what is accepted in society and what is not, with the help of the media. With his cultivation theory, Georg Gerbner studied how people's fear was influenced by watching crime on television (TV) (see Gerbner, 1972, 1973; Gerbner & Gross, 1976; Gerbner, Gross, Signorielli, Morgan & Jackson-Beeck, 1979; Gerbner et. al., 1980) and the theory is based on the fact that TV has taken such a central place in our everyday lives that it dominates our symbolic environment. Cultivation theory poses that in the long run the media alter our perceptions of reality to become a stereotyped, distorted and selective view of reality. This leads to our tendency to overestimate crime rates and the risk of becoming victims of crime and violence, and thereby we become more frightened. The theory has been criticised for failing in determining the causal direction and not taking into consideration personal experiences of crime (Doob & MacDonald, 1979). The latter point leads to the revision of the resonance theory (Gerbner, Gross, Morgan & Signorielli, 1994) positing that the double exposure of violence or crime, in the neighbourhood as well as on television, could lead to the overestimation of becoming a victim of crime or viewing the world as a scary place. Other parts of cultivation theory have studied the relationship between media coverage of crime and people's perceptions of crime, with some researchers positing strong correlations (Davis, 1952) while others positing none (Roshier, 1971, 1973; Croll, 1974). In this study, cultivation effects are investigated on the one hand due to the experiences of crime and violence in the media; on the other hand cultivation effects are investigated as correlated to expressions of Mean World Syndrome. This is done in terms of the respondents' perceptions of the crime reporting by the media at the respective level. Here respondents that perceive crime as being reported in an understated way are interpreted as perceiving the world as even scarier than in the news.

2.2 Resonance theory

When cultivation effects could not be accounted for when controlling demographic factors, or when consideration was taken to the statistics of crime (see Doob & MacDonald, 1979), other attempts were made to replicate Gerbners results, without positive results. After this critique Gerbner altered his theory into resonance theory that posed that cultivation effects occur foremost among those respondents whose social experiences are in line with the scary world that is portrayed in the media.

The overall notion is that if someone has personally become a victim of crime, s/he will notice to a larger extent media content that connotes the similar type of experience. Resonance theory poses that the influence of the media depends on what we have experiences of, therefore the more experiences we have of threats and risks, the more influenced we are by media reporting on these risks. In this study, resonance theory is used to understand how the reporting of threats and risk in the media can contribute to Mean World Syndrome and to limitations in the freedom of movement.

2.3 Enhancement effects

A more general theory on how research on threats and risk can enhance risk perceptions is Kaspersson's (Kasperson, Renn, Slovic, Halina, Emel, Goble, Kasperson & Ratick, 1988) theory on social amplification, which focuses on how we remember and try to explain how information about threats and risks are spread in society. The media, but also people in general, can act as nodes of amplification. Through the social networks of the individual, knowledge of threats and consequences of risks can be toned down, excluded or enhanced (see Kasperson, Renn, Slovic, Kasperson & Emani, 1989; Kasperson & Kasperson, 1996, p. 97), either through the character of the reporting, but also through the continuity of the occurrence of the risks. Within our study Kasperson's thesis is of importance as it is based on the assumption that memory is not only a psychological process in which we recover information, but that this recovery is dependent on various social structures that help us remember. If we ourselves do not remember, there are others who help us. Media, but also friends and acquaintances, etc., may serve as amplification stations for us to remember. People, but also the media, can opt out, tone down or reinforce certain consequences. This can be done by account of the nature or continuity in the presence of risks. This means that it is not only through our personal experience of threats and risks that we may remember unpleasant events, but also through our social and mediated experience of threats and risks.

3. Methodology and survey data

The results presented here are based on data from two set of surveys. The one concerning the urban level stems from a regional survey (from Western-SOM)[1] performed in 2003-2004 on approximately 3000 inhabitants in the local region of Gothenburg. I used data from the inhabitants in the municipality of Gothenburg with a sample of approximately 1000 persons each year. The answering frequency is approximately 65 and analyses are routinely made on why people don't answer – the most common factor being that people living in low-income areas tend not to answer questions, but no scaling of the different groups have been made. The questionnaire was in Swedish. The results have been compared to those proceeding from a local survey concerning the suburban level performed in 2011 by one of the major public housing companies in Sweden, with the objective of measuring safety in the neighbourhoods of approximately 2500 tenants. The buildings are located in different areas of the city and are composed of rented apartment blocks. In other words, there are socio-economic differences in the populations of the two studies to consider in the interpretation of the results. Another aspect to consider is whether it is possible to compare results from 2003 and 2011 at all. What we know from studying the sensations of fear and insecurity in the same city is that the results at the urban level have been very stable from the first measurements in 2001 to the last in 2007. This should also be considered when interpreting the results.

Although the two surveys have different purposes and ask different types of questions, the same frame was applied to the main dependent variable (degrees of freedom of movement) and to the two main independent variables. The dependent variable seeks to grasp the degree of limitations in the freedom of movement in public spaces during both night and day. The question was: *Are there any places or areas in your neighbourhood that you avoid due to worry, fear or insecurity?* The question was asked both concerning daytime and after dark, and alternative answers are given on an 11-grade scale from 0 (no places) to 10 (a whole lot of places). The questions and alternative answers on the experiences of different kinds of threats were: *during the last 12 months, in the area that you live in, have you been exposed to, seen, heard or read about* (more than one alternative can be marked on each row): Self been

[1] The SOM-Institute at the University of Gothenburg is jointly managed by the department of Journalism, Media and Communication, the Department of Political Science and the Centre for Public Sector Research at the University of Gothenburg.

exposed to; Self seen; Heard of friends/acquaintances; Read/seen/heard about in the press, TV, radio; No, neither. The alternative threats have varied between the two surveys, but the following alternatives have been asked about in both: Theft; Burglary; Vandalism and graffiti; Serious physical violence; People that behave violently/threatening; Serious threats that made you afraid; Presence of drug abusers; Littering. The questions and alternative answers on the perceptions of crime coverage in the surveys was the following: *Do you consider crime coverage in the radio, TV and newspapers to be an accurate representation of the actual crime rate in your neighbourhood, or do you think the media covers crime as lower or as higher than it actually is*? The alternative answers in the survey were three: Crime is portrayed as *lower* than it actually is. The media give an accurate image of the crime rates; Crime is portrayed as *higher* than it actually is. In the SOM-survey the question was framed on a more general level.

4. Freedom of movement in public places and its limitations

In this study, freedom of movement can theoretically be categorized in four distinct groups but empirically in three, since only a few respondents enjoy freedom solely after dark. The majority of the inhabitants, more than half, enjoys freedom of movement and avoids very few places both during daytime and after dark (table 1). One third of the inhabitants enjoy freedom of movement during daytime, but feels limited (avoids many places) after dark. And the most limited group, almost one sixth of the inhabitants, feels limited in their freedom of movement both during daytime and after dark.

When re-analysing the data from 2003 at the urban level, there are no differences between the regional and urban level in the freedom of movement and its limitations–as could be expected from the previous study of the sensations of fear and insecurity (Sandstig, 2010b). If we assume that it is possible to compare the levels in the freedom of movement from 2003 on the regional/urban level with the suburban data from 2011, the analysis shows that there are differences in the freedom of movement. One fourth (25 percent) of the inhabitants on the regional/urban level enjoy the freedom of movement irrespective of the time of day or night. This is less than half of the amount of subjects that enjoy freedom of movement at the suburban level. On the urban/regional level, the majority (49 percent) of the inhabitants enjoys freedom of movement during daytime but is limited after dark (as compared to the approximately one third of the inhabitants at the suburban level). When it comes to the most vulnerable, the inhabitants that limit their freedom of movement during both daytime and after dark, a little more than half (26 percent) are inhabitants at the urban level. The corresponding share on the suburban level is lower.

Table 1 - Freedom of movement in public places at suburban level 2011 and at urban level 2003 (per cent).

	Gender		Age				Education		
	Male	Female	15-29	30-49	50-64	65-85	Low	High	Total
Day and night	61(41)	44(12)	46(24)	53(25)	53(25)	51(25)	50(25)	54(24)	52(25)
Day not night	24(40)	41(56)	39(51)	32(52)	33(48)	34(42)	31(47)	37(51)	33(49)
Not day not night	15(19)	15(32)	15(25)	15(23)	14(27)	15(33)	19(28)	10(25)	15(26)
Total per cent	100	100	100	100	100	100	100	100	100
N	696 (439)	843 (575)	240 (251)	623 (359)	402 (238)	269 (165)	725 (429)	790 (560)	1573 (1013)

If we look closer at what characterizes the basic socio-demographics of the three types of groups that enjoy freedom of movement, men are the group that enjoys the most freedom of movement during both daytime and after dark at the suburban level, while women mostly enjoy freedom of movement during daytime but not after dark (table 1). The group that is most restricted in its freedom of movement (in terms of avoiding many places during both daytime and after dark) is composed by the inhabitants with a lower degree of education. At the suburban level no questions were asked about other socio-economic factors like disposable household income. What we know from the study on the urban level is that there is a pattern where socio-economically weaker groups experienced sensations of fear in public spaces to a larger extent (the reason we ask the respondents to what extent they avoided places), but that these differences were not significant.

Characteristically, the respondents who avoid few places are men, and the respondents who avoid many places are women or older citizens (table

1). The older the respondent, the lesser his/her freedom of movement in urban as well as suburban places. There are no differences in these characteristics between the two levels. However, the differences between men and women and between different age groups are greater at the urban level than on the suburban. This meets our expectations, since the restrictions in the freedom of movement were lower in the suburban areas.

In the public debate there is a common understanding that people living in socio-economically constrained areas are also the ones feeling more unsafe or unsecure. There is also a high correlation between not feeling safe and secure at the suburban level and avoiding many places. In the case of Gothenburg, however, even if such a pattern exists (Sandstig, 2010a), the differences were not significant. However, even though the inhabitants enjoy more freedom of movement at the suburban level irrespective of their level of education, it is worth noting that the differences in the avoidance of public places after dark are greater at the suburban level than at the urban level.

5. Possible influences of independent variables

5.1 Personal, social and mediated experiences of violence and crime

The question of interest in this paper is to elaborate on cultivation effects at the suburban level. They are on the one hand related to media experiences of crime, violence and threats and on the other to respondents' perceptions of media coverage of the same phenomena. In this paragraph I take a closer look at the respondents' primary and secondary experiences of crime, violence and threats in their own neighbourhood to further analyse whether cultivation effects due to media experiences occur at the suburban level and if so, how.

There are mainly two kinds of differences in experiences of crime, violence and threats at the suburban level and on a broader urban level. On the suburban level, the respondents have directly seen others become victims of crime or threat to a higher degree than at the urban level; and they have read, listened to or watched crime, violence or threats concerning their own neighbourhood in the media to a lower degree (table 2). We know from a previous study (Sandstig, 2010a; Sandstig 2010b) that at the urban level the importance of having seen others become victims of crime, violence or threat has the same independent effect on the sensations of insecurity and fear (mainly when the subject is alone in desolate spaces) of personally becoming a victim. The importance (in terms of the magnitude of the effects) of personal experience on the freedom of movement at the suburban level and the importance (in a similar sense) of cultivation effects due to mediated experiences enhancing personal experiences remain to be examined.

Table 2 - Experiences of violence/threats, theft/vandalism and littering and experiences of alcohol/drug abuse (per cent 2011 and average per cent 2003/2004).

	Type of experience				
Type of threat	**Self become a victim of**	**Seen others become victims of**	**Total personal experience of**	**Social experience of**	**Media experience of**
Violence/threats	7 (8*)	17 (27)	22 (30)	22 (24)	32 (81**)
Burglary/theft	14 (23*)	38 (10*)	47 (36)	31 (24)	26 (56**)
Littering/abuse	14 (-*)	57 (67**)	63 (67**)	12 (9**)	8 (26**)
At least some	25 (27*)	65 (44)	70 (55)	42 (37)	41 (88**)
	1754	1754	1754	1754	1754
	(1030)	(2065)	(2065)	(2065)	(1035)

Commentary: Data from 2003 within brackets, N 2011=1754; N 2003/2004=2065; *) N 2003=1030; **) N 2004=1035

5.2 Perceptions of crime coverage in the media

There is a difference in the primary perception between the urban and the suburban level. At the suburban level the perception that the media coverage is accurate is the most common, while the most common at the urban level was that media coverage of crime was understating crime and insecurity (table 3).

At the urban level respondents with lower degrees of education perceived media coverage to be understated, meaning that the world outside was even more threatening than in the media; the elderly perceived that the reporting was accurate and the young and well educated perceived the coverage to be overstated. At the suburban level the pattern was the same, but not as obvious. Like at the urban level, it was mainly the respondents with lower degrees of education who perceived media coverage of crime in their neighbourhood as understated, but not to the same extent. When it comes to perceptions of accurate media coverage, age differences were missing at the suburban level; however, it was mostly the young and well educated who perceived the media coverage to be overstated.

Table 3 - Perceptions of media coverage at the suburban level 2011 and in general 2003 (per cent)

Perceptions of media coverage of crime	Gender		Age				Education		Total
	Male	Female	15-29	30-49	50-64	65-85	Low	High	
Understated	25 (43)	18 (39)	17 (36)	20 (41)	22 (44)	24 (42)	25 (48)	18 (34)	21 (41)
Accurate	50 (29)	58 (34)	53 (29)	56 (32)	52 (31)	55 (36)	52 (32)	56 (32)	54 (32)
Overstated	25 (28)	24 (27)	30 (35)	24 (27)	26 (25)	21 (22)	23 (20)	26 (34)	25 (30)
Total per cent	100	100	100	100	100	100	100	100	100
N	693 (432)	818 (562)	237 (242)	595 (359)	397 (230)	277 (163)	728 (415)	753 (553)	1537 (994)

Commentary: Data from 2003 within brackets.

6. Cultivation, resonance and enhancement effects

6.1 Related to mediated and personal experience of crime and violence

While the suburban level also shows some cultivation effects (due to both personal and mediated experience) on the limitation of the freedom of movement, they are lower than at the urban level. At the urban level, among those with both personal and mediated experiences of violence and threats, approximately 43 per cent enjoy freedom of movement both during daytime and after dark (table 4). The corresponding share at the suburban level was almost the same (40 per cent). The independent effects of these cultivation effects remain to be analysed at the end of this paper.

Table 4 - Percentages who limit their freedom of movement in public places at urban and suburban level among those who have media experience of crime, violence or threat (per cent 2011 and per cent 2003)

Freedom of movement in public places	Mediated experience but lack of personal	Both personal and mediated experience	Social experience but lack of mediated
Day and night	43(25)	28(10)	37(23*)
Day not night	40(51)	47(46)	38(46*)
Not day not night	16(24)	25(44)	25(31*)
Total per cent	100	100	100
N	391(699)	137(115)	201(26*)

Commentary: Data from 2003 within brackets. * Please note the low amount of respondents. 2003

If we consider the interpretation of those exposed both on a personal level as well as through the media, the avoidance of public places on the suburban and urban level can be interpreted through the use of resonance theory (Gerbner, 1994) or the criticism (see Doob & MacDonald, 1979) directed against cultivation theory. Resonance theory illustrates how personal experience of crime may increase media influence. Research has shown that cultivation effects were limited to those living in crime-burdened areas. The interpretation is that those who received a double dose of crime were most afraid of being exposed to crime. The results within the framework of resonance theory have been interpreted as the mass media and personal experience rather reinforcing each other (Doob & MacDonald 1979, Gerbner et al, 1994). But my previous results show that first-hand experience of being a crime victim is not the only aspect relevant to the sensation of fear and insecurity in public spaces: there is also relevance in having seen others become victims of crime and even threats and risks in general. This suggests that it is what people have knowledge of, or know about, that is crucial to a feeling of fear, not only by personally becoming a victim but also by having seen others become victims of crime or threats and risks, or by reading about it in newspapers or taking note of it via the media. Resonance theory in general states that media influence occurs because we pay greater attention to what we also have experience of, so the more experience we have of crime, violence or threats, the more influenced we will become by the media coverage of these events. So lower cultivation effects on the avoidance of places at the suburban level could be explained with lower degrees of personal and/or mediated experiences of violence and threats at the suburban level (see table 4).

6.2 Enhancement effects of mediated, personal and social experience of crime and violence

In our previous study on the urban level, the mediated experience did not have any independent effect of its own on the sensations of fear and insecurity. However, it did enhance not only personal experiences but also social experiences. Those that had become victims themselves had seen others become victims of crime, violence or threats or had heard about crime, violence and threats through friends and acquaintances and had mediated experiences of the same, experienced the sensation of fear and insecurity in public spaces to a higher degree. In fact, the more types of experiences of crime, violence and threats the respondents had, the more fear and insecurity they had experienced in public spaces.

The pattern of the avoidance of public places is the same at the suburban level: the more types of experiences you have of violence and threats in or of your neighbourhood, the more likely it becomes to avoid many places (table 5). There are differences between the suburban and urban levels, in that those lacking any type of experience are freer to move at the suburban level and those having all three kinds of experiences (personal, social and mediated) to a greater extent avoid many public spaces at the suburban level.

A reasonable explanation for how the three routes of experience can reinforce each other in terms of influence on the avoidance of public places after dark is through Kasperson's (1988) theory of social amplification that focuses on how we remember and try to explain how information about threats and risks is spread in society. Here memory is not solely seen as an individual psychological process. Our remembering is also dependent on certain social structures that help us remember. Here friends and acquaintances, as well as the media, can act as nodes of recovering and either amplify or tone down threats and risk perceptions. In other words, through our own experiences we can remember unpleasant events but also our social experiences of such events transmitted through friends and acquaintances as well as the media. The increasing complexity of dealing with these experiences can lead to a lower degree of control over one's environment that also has been shown to increase fear and insecurity.

Table 5, Percentages who limit their freedom of movement in public places on urban and suburban level dependent on the degree of experiences of crime, violence or threat (per cent 2011 and 2003).

Freedom of movement in public places	Lack of experience	Had one type of experience	Had two types of experiences	Had three types of experiences
Day and night	68 (35)	43 (27)	30 (13)	25 (10)
Day not night	24 (45)	40 (51)	42 (49)	50 (43)
Not day not night	8 (20)	17 (22)	28 (38)	25 (47)
Total per cent	100	100	100	100
N=1675 (1001)	724 (141)	534 (617)	263 (178)	52 (58)

Commentary: Data from 2003 within brackets.

6.3 Cultivation effects due to the perception of crime reporting

As previously shown, at the urban level, those who perceive that media coverage of crime is understated are those who also to a larger extent experienced sensations of fear and insecurity in public places. However, perceiving media coverage to be understated, or believing that the world outside is even more threatening than in the media, does not lead to higher restrictions in the freedom of movement. On the contrary, the respondents who perceive the media coverage of crime to be understated feel so to a lesser extent than the respondents perceiving media coverage as accurate or overstated public places due to worries, fear or insecurity (table 6).

When it comes to perceptions of media coverage of crime in the neighbourhood, the respondents who perceive media coverage to be accurate avoid the most places (table 6). One explanation for these differences is that fewer people living in these suburban areas read traditional news media and, as cultivations theory poses, the lack of influence from the media has not distorted their view on reality.

Table 6 - Avoidance of places at the suburban level when alone due to worry, fear or insecurity dependent on perceptions of media coverage of crime in one's own neighbourhood dependent on (per cent)

Freedom of movement in public places	Understated	Accurate	Overstated
Day and night	40(18)	51(26)	61(29)
Day not night	39(48)	33(51)	31(51)
Not day not night	21(34)	16(23)	8(20)
Total per cent	100	100	100
N	285 (395)	786 (309)	352 (270)

Commentary: Data from 2003 within brackets.

To further analyse what factors have the greatest independent effect on the urban and suburban level, a multiple linear regression analysis was used. The results imply that personal experience (having seen others become victims or having first-hand experience of victimhood) of violence and threats has the greatest independent effect at the suburban level in limiting the freedom of movement after dark and in general. Mediated and social experiences (having heard friends or acquaintances become victims of crime) have the second greatest independent effect on the freedom of movement in general, while gender has the second greatest independent effect on the freedom of movement after dark (table 7). However, during daytime it is the level of education that has the greatest independent effect at the suburban level, followed by personal experience. If we look closer at the limitations on the freedom of movement during both daytime and after dark, the pattern is the same.

Compared to the urban level, the correlations show some differences. At the urban level, gender has first and foremost an independent effect on the freedom of movement in general, and in most of the other cases (table 8) not personal experience at the suburban level. However, personal experiences and perception of crime coverage in the media also matter in general: the higher the personal experience of the subject, and the more understated the perception of media coverage, the higher the limitations in the freedom of movement.

Table 7 - The effect of experiences (personal, social and media) of violence, socio-demographic factors and perceptions of media coverage of crime on the freedom of movement at the suburban level (2011) (standardized regression coefficient, b-value and significance, p-value)

	After dark Multivariate effect (stand. B-value)	Daytime Multivariate effect (stand. B-value)	Degrees of freed. Multivariate effect (stand. B-value)	Free day, free night Multivariate effect (stand. B-value)	Free day, limited night Multivariate effect (stand. B-value)	Limited day, limited night Multivariate effect (stand. B-value)
Personal exp. of violence	$0,22^{***}$	$0,12^{***}$	$0,19^{***}$	$-0,20^{***}$	$0,20^{***}$	$0,12^{***}$
Social exp. of violence	$0,12^{***}$	$0,08^{**}$	$0,14^{***}$	$-0,12^{***}$	0,05	$0,11^{***}$
Media exp. of violence	$0,13^{***}$	$0,06^{**}$	$0,15^{***}$	$-0,16^{***}$	$0,10^{***}$	$0,08^{**}$
Gender	$0,18^{***}$	0,07	$0,13^{***}$	$-0,18^{***}$	$0,18^{***}$	0,01
Age	$0,05^{***}$	-0,05	$0,02^{***}$	-0,03	0,03	0,00
Educa-tion	$-0,10^{***}$	$-0,14^{***}$	$-0,12^{***}$	$0,07^{*}$	0,04	$-0,15^{***}$
Percep. crime cover-age	$-0,11^{***}$	$-0,09^{***}$	$-0,11^{***}$	$0,10^{***}$	-0,04	$-0,08^{**}$
R2 (%)	0,15	0,06	0,13	0,14	0,06	-0,06
Total	1401	1372	1341	1341	1341	1341

Commentary: N=1401, R=1401, * $p<.05$; ** $p<.01$; *** $p<.001$.

Table 8 - The effect of experiences (personal, social and media) of violence, socio-demographic factors and perceptions of media coverage of crime on the freedom of movement at the urban level (2003) (standardized regression coefficient, b-value and significance, p-value)

	After dark Multivariate effect (stand. B-value)	Daytime Multivariate effect (stand. B-value)	Degrees of freed. Multivariate effect (stand. B-value)	Free day, free night Multivariate effect (stand. B-value)	Free day, l limited night Multivariate effect (stand. B-value)	Limited day, limited night Multivariate effect (stand. B-value)
Personal exp. of violence	0,08**	0,13***	0,16***	-0,10 ***	-0,05	0,15***
Social exp. of violence	0,10**	0,09**	0,10**	-0,07*	-0,02	0,09**
Media exp. of violence	0,03	-0,01	0,04	-0,04	-0,01	0,03
Gender	0,36***	0,16***	0,30***	-0,35 ***	0,16***	0,15***
Age	0,08**	0,11***	0,08**	-0,03	-0,07*	0,10***
Education	-0,01	-0,02	-0,01	0,00	0,01	-0,02
Percep. crime coverage	-0,14***	-0,11***	-0,13 ***	0,09**	0,02	0,08
R2 (%)	0,17	0,07	0,15	0,15	0,03	0,08
Total	946	946	942	942	942	942

Commentary: N=1401, R=1401, * $p<.05$; ** $p<.01$; *** $p<.001$.

These patters resemble the pattern in the limitations in the freedom of movement during daytime at the urban level and those of the people who enjoy freedom of movement both during daytime and after dark. In the limitation of freedom of movement the perceptions of media coverage also matter, especially after dark, as opposed to the suburban level. The main conclusion is that the perception of media coverage of crime in general has greater independent effects at the urban than at the suburban level and especially after dark. At the suburban level, personal and other experiences have greater independent effects, and if we look at the differences between groups, the level of education is the most limiting factor in the freedom of movement during daytime and for those avoiding many places during both daytime and after dark (at the suburban level as opposed to the urban level). Whether these differences are the methodological outcome of the different populations or actual differences between the suburban and urban levels remains to be further studied.

Acknowledgements

This paper is based on a study partly funded and performed in collaboration with Förvaltnings AB Framtiden, a public housing company entirely owned by the municipality of Gothenburg, Sweden. The author wishes to thank Bo Säljö, former head of security at Förvaltnings AB Framtiden and Peter Lundstedt, CEO at Familjebostäder for their support.

References

Bauer, M. W. (2005). Distinguishing red and green biotechnology: Cultivation effects of the elite press. *International Journal of Public Opinion Research*, 17 (1), 63-89.

Bradley, S. D. (2007). Neural network simulations support heuristic processing model of cultivation effects. *Media Psychology*, 10 (3), 449-469.

Busselle, R. W. (2001). Television exposure, perceived realism, and exemplar accessibility in the social judgement process. *Media Psychology*, 3 (1), 43-67.

Cheung, C. K., & Chan, C. F. (1996). Television viewing and mean world value in Hong Kong's adolescents. *Social behavior and Personality*, 24 (4), 351-364.

Croll, P. (1974). *The deviant image*. Paper presented at British Sociological Association Mass Communication Study Group.

Davis, F. (1952). Crime news in Colorado newspapers. *American Journal of Sociology*, 57, 325-330.

Doob, A. M., & McDonald, G. E. (1979). Television viewing and fear of victimization: Is the relationship casual? *Journal of Personality and Social Psychology*, 37, 170-179.

Dudo, A., Brossard. D. Shanahan, J. Scheufele, D.A., Morgan, M., & Signorielli, N. (2011). Science on Television in the 21st Century: Recent Trends in Portrayals and Their Contributions to Public Attitudes Toward Science. *Communication Research*, 38 (6), 754-777.

Gerbner, G. (1972). Communication and social environment. *Scientific American*, 227, 235-260.

—. (1973). Cultural indicators: The third voice. In G. Gerbner, L. Gross & W. H. Melody (Eds.), *Communications Technology and Social Policy*. New York: John Wiley.

Gerbner, G. (1980). Death in Prime Time: Notes on the Symbolic Functions of Dying in the Mass Media. *Annals of the American Academy of Political and Social Science*, 447, 64-70.

Gerbner, G., & Gross, L. (1976). Living with television: The violence Profile. *Journal of Communication*, 26 (2), 172-194.

Gerbner, G., Gross, L., Signorielli, N., Morgan, M., & Jackson-Beeck, M. (1979). The demonstration of power: Violence profile no. 10. *Journal of Communication*, 29 (3), 177-196.

Gerbner, G., Gross, L., Morgan, M., & Signorielli, N. (1980). The mainstreaming of America: Violence profile no. 11. *Journal of Communication*, 30 (3), 10-29.

Gerbner, G., Gross, L., Morgan, M., & Signorielli, N. (1994) Growing up with Television: The Cultivation perspective. In J. Bryant & Zillmann D. (Eds.), *Media Effects. Advances in Theory and Research*. Hillsdale NJ: Lawrence Erlbaum Associates.

Good, J. E. (2009). The Cultivation, Mainstreaming, and Cognitive Processing of Environmentalists Watching Television. *Environmental Communication–A Journal of Nature and Culture*, 3 (3), 279-297.

Hale, C. (1996). Fear of Crime: A Review of the Literature. *International Review of Victimology*, 4, 79-150.

Heber, A. (2005). *Var rädd om dig: En litteraturöversikt om rädslan för brott*. Stockholm: Kriminologiska institutionen, Stockholms universitet.

Hetsroni, A. (2010). When the wind changes direction: The impacts of content shift on the cultivation effect. *Communications–European Journal of Communication Research*, 35 (4), 439-460.

Harvey, D. (2001). *Spaces of Capital: Towards a Critical Geography*. Edinburgh: Edinburgh University Press.
Harvey, D. (2009). *Cosmopolitanism and the Geographies of Freedom*. New York: Columbia University Press.
Kasperson, R. E., Renn, O., Slovic, P., Halina, S.B. Emel, J., Goble, R., Kasperson, J. X., & Ratick, S. (1988) The social amplification of risk: a conceptual framework. *Risk Analysis*, 8 (2), 177-188.
Kasperson, R. E., Renn, O., Slovic, P., Kasperson, J. X., & Emani, S. (1989). Social amplification of risk: the media and public response. In R. G. Post (Ed.), *Waste Management '89: Waste Processing, Transportation, Storage, and Disposal—Vol. 1: High-level Waste and General Interest* (pp. 131-135). Tucson: University of Arizona Press.
Kasperson, R. E., & Kasperson, J. X. (1996). The social amplification and attenuation of risk. In H. Kunreuther & P. Slovic (Eds.), *Challenges in Risk Assessment and Risk Management* (pp. 94-105). The Annals of the American Academy of Political and Social Science, 545.
Lee, C., & Niederdeppe, J. (2011). Genre-Specific Cultivation effects: Lagged Association Between Overall TV Viewing, Local TV News Viewing and Fatalistic beliefs About Cancer Prevention. *Communication Research*, 38 (6), 731-753.
Massey, D. (1994). *Space, Place and Gender*. Oxford: Polity Press.
Mitchell, D. (2004). The Right to the City: Social Justice and Right to the City. *Economic Geography*, 80 (4), 407-408.
Morgan, M., & Shanahan, J. (1997). Two decades of Cultivation Research: An Appraisal and a Meta-Analysis. In B. Burleson (Ed.), *Communication Yearbook 20* (pp. 1-45). Thousand Oaks, CA: Sage.
Nabi, R. L., & Riddle, K. (2008). Personality traits, television viewing and the cultivation effect. *Journal of Broadcasting & Electronc Media*, 52 (3), 327-348.
Romer, D., Jamieson, K. H., & Aday, S. (2003). Television news and the cultivation of fear of crime. *Journal of Communication*, 53 (1), 88-104.
Roshier, R. (1971). Crime and the press. *New Society*, 486, 502-506.
—. (1973). The selection of crime news by the press. In S. Cohen & J. Young (Eds.), *The manufacture of news*. London: Constable.
Sandstig, G. (2010a). *Otrygghetens landskap. En kartläggning av otryggheten i stadsrummet och en analys av bakomliggande orsaker, med fokus på mediernas roll*. Institutionen för Journalistik, medier och kommunikation, Göteborgs Universitet.
—. (2010b). Media influences and effects on experiences of uncertainty and fear in urban public spaces. *Journal of US-China Public Administration*, 7 (12), 66-84.

—. (2013). Backabornas samhällsanda 1994-2011. SOM-rapport nr 2013:2. Göteborg: SOM-Institutet, University of Gothenburg.

Shrum, L. J. (1993). Processes and Effects in the Construction of Social-Reality–Construct Accessibility as an explanatory Variable. *Communication Research*, 20 (3), 436-471.

—. (1996). Psychological processes underlying cultivation effects–Further tests of construct accessibility. *Human Communication Research*, 22 (4), 94-120.

—. (1997). The role of source confusion in cultivation effects may depend on processing strategy–A comment on Mares (1996). *Human Communication Research*, 24 (2), 349-358.

—. (2001a). Processing strategy moderates the cultivation effect. *Human Communication Research*, 27 (1), 94-120.

—. (2001b). Mainstreaming, resonance, and impersonal impact–Testing moderators of the cultivation effect for estimates of crime risk. *Human Communication Research*, 27 (2), 187-215.

—. (2007). The implications of survey method for measuring cultivation effects. *Human Communication Research*, 33 (1), 64-80.

Shrum, L. J., Lee, J., Burroughs, J. E., & Rindfleisch, A. (2011). An Online Process Model of Second-Order Cultivation Effects: How Television Cultivates Materialism and its Consequences for Life Satisfaction. *Human Communication Research*, 37 (1), 34-57.

Woo, H. J., & Dominick, J. R. (2001). Daytime television talk shows and the cultivation effect among US and international students. *Journal of Broadcasting & Electronc Media*, 45 (4), 598-614.

CHAPTER XV

DIALOGICAL NARRATIONS IN THE DIVIDED CITY OF MOSTAR

GONCA NOYAN

> *When Allah the Merciful and Compassionate first created this world, the earth was smooth and even as a finely engraved plate. That displeased the devil who envied man this gift of God. And while the earth was still just as it had come from God's hands, damp and soft as unbaked clay, he stole up and scratched the face of God's earth with his nails as much and as deeply as he could. Therefore [...] deep rivers and ravines were formed which divided one district from another and kept men apart [...] And Allah felt pity when he saw what the Accursed One had done [...] so he sent his angels to [...] spread their wings above those places and [...] men learnt from the angels of God how to build bridges, and therefore, after fountains, the greatest blessing is to build a bridge and the greatest sin to interfere with it [...].* (Ivo Andric, 1945, *The Bridge Over the Drina*).

Ivo Andric, 1962 Nobel Laureate from Bosnia-Herzegovina, in his novel "Bridge Over the Drina" narrated the story of God's creation of the world with a focus on bridges. Although Andric's story focused on a public space created by God, in real life public spaces are designed, built and used by human beings. Parks, plazas, streets and other spaces constitute the foundation of our daily social lives, as public space provides "an open and inclusive stage for social interaction, political action and cultural exchange" (Carmona, 2008, p. 4).

A pedestrian bridge is a public space, often built over a physical obstacle such as river for the purpose of connecting one side to the other. As public space, bridges are used and experienced in multiple forms by different individuals. One can use the bridge to cross from one side to the other as a part of a daily routine while commuting between home and work. The bridge can be used as a meeting point for friends, a venue for public protests, a spot to take the best touristic photo, a shelter for the homeless or a place for beggars and so on. The bridge has not only a physical existence (in material and architectural terms), but also an imaginative existence (as an idea or symbol). One can view the bridge as a marvel of architecture, an inspiration for a poem, a place to remember childhood memories or just an ordinary space seen every day.

Like Andric's famous novel, this study will also focus on a specific public space, which is the Old Bridge. The study will be guided by the following research question: "What are the narratives used about the Old Bridge in Mostar and how do they evolve throughout history?"

It is acknowledged that public space can be read as a narration about social relations (Low, 2000). People react to places in line with the meanings those places have for them (Rapoport, 1982). Analysis of a public space is important for presenting the discourse of power relations and allowing us to understand ongoing struggles in a specific place. Communication scholars seem particularly well-situated to understand the communicative role of the public space and to what extent the locals' perception of public space differs from that of global authorities.

To understand the meanings attached to a specific public space and how they affect narration, this study will use written material including but not limited to historical accounts, newspaper articles, travel descriptions, novels, poems, and advertisements. Those materials are essential to depict "how people see environments, how they feel about them, what they like or dislike about them" (Rapoport, 1982, p. 11).

This study will be grounded on Bakhtin's Dialogic Narration Theory, and assumes that narratives of the Old Bridge are not static but allow multiple interpretations over time depending on the context. My intention is to read the Old Bridge as a cultural text that facilitates the understanding of the meanings attached to a specific place both by global authorities, media and local residents of different ethnicities.

The study is articulated into: a literature review of the concepts of "public space" and of "dialogic narration"; a short account of the Old Bridge as a public space; and analysis of narratives about the Old Bridge.

1. Public Space

Public space provides “an open and inclusive stage for social interaction, political action and cultural exchange” (Carmona, 2008). Parks, plazas, streets, pedestrian bridges, monuments as public space constitute the foundation of our daily lives. Types of public space have evolved throughout history, ranging from ancient Roman forums to today’s shopping malls.

Public space has been studied by scholars from different fields including political science, philosophy, geography, urban planning, anthropology, visual arts and cultural studies. Scholars (Harvey, 2006; Low, 1996) have also studied the exclusionary aspects of public space, focusing on the spatial segregation of activities in terms of class, ethnicity and race. In anthropological scholarship, studies of public space have focused on the objectification and construction of other-ness (Gupta & Ferguson, 1992; Harvey, 2006). Excluding specific groups from a particular space and/or prohibiting specific activities within that space strengthens the experience of otherness in public space.

1.1 Political and Economic Dimensions of Public Space

Any spatial form, including public space, is a product of conflicting political and economic forces (Bourdieu, 1977; Foucault, 1975; Harvey, 1985; Low, 2000). Low (2000) argues that the design of a public space can reflect the political agency of the state. Similarly, Foucault (1975) asserts that public space enables the dominance of one group over another through surveillance. Bourdieu (1977) criticises the fact that the design of public space often reproduces the power relations of the past.

Since public space reflects the political and economic power structures within a society (Low, 2000), the discourse of a public space illuminates the larger political and economic process. Scholars (e.g., Carmona, 2008) have underlined the economic benefits of public space arguing that public space might have a positive impact on property prices, investment levels and economic performance. In line with the impact of globalization, the importance attached to public spaces has increased as both individuals and countries have attempted to present the distinctiveness and uniqueness of specific public spaces as if they were products at the market (Low, 1996). We observe that public space continuously evolves in response to global forces on both citizens and nation-states, with the commodification of public space, and the influence of power from distant political and economic forces. Public spaces are increasingly used for city branding to attract foreign direct investment and global tourism.

1.2 Social and Cultural Dimensions of Public Space

Scholars (Carr et al., 1992; Duncan, 1990; Low, 2000; Rapoport, 1982) have focused on the relation between public space and social and cultural identities by highlighting the communicative character of public space. In this sense, public space is seen as a venue for social encounter and interaction.

Throughout history, public spaces have supported the needs of communities in celebrations and rituals; thus, public spaces "symbolize the community and the larger society or culture in which [they] exist" (Carret al., 1992, p. 23). Accordingly, interpretations of public space should take into consideration the following factors: (1) the past and present existence of a space, (2) the social behaviour accommodated by the place, and (3) the symbolic and communicative aspects of the place (Low, 2000, p.47). Similarly, Duncan (1990) notes the possibility of reading public space as a communication about social relations by analysing contemporary understanding and historical accounts of a specific public space.

Lefebvre (1997) asserted that public space is socially produced and constructed. Low (1996) defines social construction of space as "the actual transformation of space–through people's social exchanges, memories, images, and daily use of the material setting–into scenes and actions that convey symbolic meaning" (p. 862).

1.3 Conflicting Character of Public Space

Social theorists have emphasized the importance of everyday activities and social relations in negotiating culture (de Certau, 1984; Foucault, 1984). Considering the fact that public space witnesses the daily activities of different groups within the community, public space enables us to understand the divisions and/or connections between different cultural groups. Furthermore, public space stands as an arena in which ethnic and cultural identities are constructed, challenged and continually negotiated.

Rapoport (1982) emphasizes that designers should take into consideration the symbols in a given culture when designing public space. Yet, in multicultural societies designing public space becomes problematic since "'symbols are neither fixed nor shared" (Rapoport, 1982, p. 45). According to Rapoport (1982), designing public space is "a process of encoding information" which is decoded by the users of public space (p. 57). Therefore, there is always a risk that users of a public space might not communicate if the code is not shared by all users (Rapoport, 1982).

Tunbridge and Ashworth (1996) assert that representations of public space do not usually provide "an ordered and widely intelligible communication," but rather act as "a city of Babel where numerous languages are being muttered or shouted together, most of which are only understood by some of the citizens [and] many are so obsolete as to be only dimly and incompletely comprehended" (p. 15).

Tunbridge and Ashworth (1996) emphasise the dialogic character of public spaces and buildings in cities arguing that their meanings produce "a language for which each individual possesses their own personal dictionary which is constantly and rapidly being replaced over time" (p. 15). For instance, a monument of a political figure in a public space might evoke different interpretations of this particular place depending on one's political views. Furthermore, the changing of names of public spaces in line with the dominant political authorities reflects the contested status of public spaces. The conflicted nature of public spaces leads to the presence of diverse tellings and thus to dialogic narration.

2. Dialogic Narration

Narrative has been used as a scholarly tool in many disciplines, including history (Wertsch, 2000; White, 1987), language (Bakhtin, 1981, 1984, 1986; Lotman, 1988, 1990), and cultural psychology (Bruner 1990; Nielsen, 1987). Many such studies, across the disciplines, have been founded upon Mikhail Bakhtin's (1981) ideas about dialogic narration.

2.1 Existence of Multiple Voices

Bakhtin (1986) emphasizes the fact that all texts are mediated and defines a language system as a cultural tool that allows "repeatable and reproducible" elements by each individual speaker (1981). In that sense, Bakhtin's perspective does not take semiotics as something neutral and lifeless; he rather asserts that all utterances are characterized by "multivoicedness" and "dialogicality" (1981, p. 337). In a sense, any speech, text or built environment is shaped by cultural, historical and social contexts.

Narratives do not exist in isolation; they occur as a part of discourse, which is characterized by dialogical and rhetorical processes (Wertsch, 2000). Each individual engages in dialogical discourse because we define ourselves and others in response to previous utterances. Bakhtin (1981) emphasizes the dialogical dynamism of cultural semiotic life through the recognition of the existence of multiple voices. Because there is no limit to the number of voices in any dialogue, dialogic processes enable multiple voices to exist and challenge each other. Bakhtin also argues that any two discourses "oriented to the same referential object within the limits of a single context cannot exist side by side without intersecting dialogically, regardless of whether they confirm, mutually supplement, or (conversely) contradict one another" (1984, p. 187). Similarly, Wertsch asserts that "the meaning and form of one narrative may be understood in terms of being a dialogic response to another, previous one, in terms of anticipating another, subsequent narrative, and so forth" (2000, p. 517).

2.2 Dialogic Narration of a Public Space

Public space can be read as a narration about social relations (Low, 2000) since people react to places consistently with the meanings those places have for them (Rapoport, 1982). Scholarly tools that can be used to analyse public spaces include "reading the landscape as a text, decoding the built environment through analogy and metaphor, and the phenomenological experience of place" (Low, 2000, p. 49).

Studying the discourse of public space, scholars have both highlighted the dialectic relationship between memory and past (Katriel, 1994) and the dialogical processes of multiple voices (Bruner & Gorfain, 1984). Katriel (1994), in her study of heritage museums in Israel, noted the coexistence of two orientations toward the past in the narratives of tour guides and of museum displays. She defines heritage museums as "cultural sites of articulation of this very basic tension of history and memory" (p. 16). On the other hand, Low (2000) defines the social production of space as a dialogical process considering the existence of "a high degree of conflict and contestation" (p. 37). In a similar tone, Bruner and Gorfain (1984) assert that "no story is 'a' story or 'the' story but rather a dialogic process of many historically situated particular tellings" (p. 57). Focusing on the story of the mountain fortress of Masada, the authors describe how narratives of a specific place evolve over time in response to changing circumstances and reflect the socio-ideological conflicts that are in play within the society at a given cultural moment. Bruner and Gorfain (1984) describe the various tellings and interpretations of the same story as a process of dialogic narration. Dialogic narration has three domains: the *intrinsic*, the *historical* and the *experiential*. In line with those domains, narratives about a public space respond and interact with each other (the *intrinsic*), with communities and their histories (the *historical*), and with the individuals in those communities (the *experiential*).

In contrast to Katriel's (1994) binary approach to discourse of public space, Bruner and Gorfain (1984) present a broader view that includes multiple voices in discourses of public space. Because Mostar's Old Bridge is located in a multiethnic society and partially under the jurisdiction of international authorities, an approach emphasising the existence and negotiation of diverse voices is more suitable for the analysis of narratives about the Old Bridge.

The Old Bridge in Mostar represents a unique public space since it can be interpreted both as a bridge dividing the east-Bosnian Muslim side from the west-Croat Catholic side, and as a connecting bridge considering the intermarriages and presence of different ethnic and religious groups on both geographical sides. In this framework, the Old Bridge provides a microcosm of the society where different identities encounter and negotiate in the public space; the Old Bridge in Mostar has been simultaneously a tourist attraction, a national monument of Bosnia-Herzegovina, a UNESCO world heritage, a subject of poetry, a symbol of multiculturalism and even a battlefield. Each of these occasions presents different narrative constructions of the Old Bridge. Understanding any given narration not as something stable but as a flexible, organic process, this study will look at how narratives about the Old Bridge have evolved and changed. Before focusing on the narratives, the next section presents a short account of the Old Bridge in Mostar as a public space.

3. The Old Bridge in Mostar as a Public Space

The Old Bridge is a 16th century pedestrian bridge that stands across the river Neretva and connects the two sides of the city of Mostar in Bosnia and Herzegovina. The Old Bridge was constructed in 1566 as

Mostar become an administrative centre of the region under the Ottoman rule. The city of Mostar was named after the Old Bridge as a reference to the "bridgekeepers" that guarded the bridge. (Mostar means "bridgekeepers" in Bosnian language.)

Before the devastating Bosnian War, the population of Mostar was composed of one-third Muslims, one-third Croatians and one-fifth Serbians (Grodach, 2002). The partition of Bosnia and Herzegovina into quasi-ethnic regions after the Bosnian War changed the ethnic structures of the city. Serbians displaced or migrated to the Republika Srpska (dominated by Serbs) while Bosnians and Croats stayed in Mostar which belongs to the Federation of Bosnia and Herzegovina (shared by Croats and Bosnians). Currently, Mostar is mostly populated by Muslims and Croatians, and the Old Bridge connects the residents of East Mostar with the residents of West Mostar.

Bosnians, Croats and Serbs had lived peacefully in the city of Mostar until the Bosnian War of 1992-1995. The bridge was destroyed by the shelling of the Bosnian Croat Military during a few hours on the 9th of November 1993.

The destruction of the Old Bridge in Mostar disappointed the Bosnians who had hoped for the reconstruction of a multiethnic society in Bosnia and Herzegovina (Donia & Fine, 1994; Malcolm, 2002; Pašić, 2006). The partition of Bosnia and Herzegovina into quasi-ethnic regions after the war changed the composition of population and the ethnic structures of the city. Serbians were displaced or migrated to the Serb-dominated Republika Srpska, while Bosniaks and Croats stayed in Mostar under the Federation of Bosnia and Herzegovina (shared by Croats and Bosniaks).

After the war, the city of Mostar was segregated into a Muslim area on the East side of the Neretva River and a more developed Croat side on the West (Schuman, 2004, p. 126). Although Serbs who migrated to the Serb-dominated parts of the country during the war have recently started to return to the city, Mostar today is largely a Bosniak-Croat city. According to the Federation of Bosnia Herzegovina's Office of Statistics (2009), the population of Mostar was 111,186 in 2009. There is no information about the ethnic composition of the city; ethnic and religious questions are not included in the census due to continuing sensitivities.

Initial efforts by international organisations like the EU to reintegrate Mostar did not succeed due to strong resistance from locals. Mostar became more a divided city organised into six municipalities and a central zone with equal shares of authority. Furthermore, nearly all of the local administrative structures, including healthcare, public education, infrastructure and communication systems were divided. Political and institutional divisions have had repercussions on the daily lives of residents as well.

Following the signing of the Dayton Agreement which ended the war, the restoration project of the Old Bridge was initiated and overseen by a scientific committee established by UNESCO. The project was funded by international financial aid, particularly from Italy, Turkey, France and Netherlands. The rebuilt Old Bridge opened with a ceremony on July 23, 2004.

The Old Bridge was designated as a National Monument of Bosnia and Herzegovina on July 12th 2004. Since July 2005, the Old Bridge and its surrounding area are included into UNESCO's World Heritage List. UNESCO's World Heritage List currently includes 890 properties (689 cultural, 176 natural, and 25 mixed properties) as representing the cultural and natural heritage (UNESCO, 2010). The properties are chosen by the World Heritage Committee, the main institutional body responsible for the implementation of the World Heritage Convention which includes the criteria and guidelines for the inscription of the properties (UNESCO, 2010). The list includes unique places such as Egypt's Great Pyramids, Turkey's Archaeological Site of Troy, China's Great Wall and the United States' Grand Canyon National Park.

UNESCO's World Heritage List is important for being universal; the sites included in the list "belong to all the peoples of the world, irrespective of the territory on which they are located" (UNESCO, 2010).

UNESCO substantiates the inclusion of the Old Bridge by defining the bridge and the surrounding area as "an exceptional and universal symbol of coexistence of communities from diverse cultural, ethnic and religious backgrounds" (UNESCO, 2010). Currently, the Old Bridge area presents a true example of multicultural urban settlement thanks to its architectural features including pre-Ottoman, Ottoman, Mediterranean and European traces, and welcomes tourists from all around the world to see the Old Bridge (UNESCO, 2010).

The Old Bridge represents a unique public space since it might be interpreted both as a bridge dividing the east-Bosnian side from the west-Croat side and as a connecting bridge considering the intermarriages and presence of different ethnic and religious groups on both sides of the bridge. In this framework, the Old Bridge provides a microcosm of a society where different identities encounter and negotiate in the public space.

As the brief historical account given above denotes, the Old Bridge has been simultaneously a tourist attraction, a national monument of Bosnia-Herzegovina, a world heritage site approved by UNESCO, a subject for poems, a symbol of multiculturalism and even a battlefield. On all those occasions, the narration of the Old Bridge presents a different telling. Shall the Old Bridge be a subject of multiculturalism or a sign of cultural hatred? Which one shall we tell while defining the Old Bridge?

According to UNESCO, the Old Bridge is "an exceptional and universal symbol of coexistence of communities from diverse cultural, ethnic and religious backgrounds" (UNESCO, 2010). In this version, the Old Bridge represents the ideal of ethnic amity and cultural diversity.

How about this version, an excerpt from a letter written to the editor of a local daily newspaper (Dube, 1993):

> The destruction of one of the world's most exquisite bridges–the Old Bridge of Mostar, which stood for 427 years–is undoubtedly the culminating act of barbarism in the Yugoslav war. How could the civilized world allow this to happen? It had 2 1/2 years to stop the war. Until Nov. 8 the bridge was still standing, although damaged by months of Croatian shelling. On Nov. 9 the forces of barbarism finally triumphed in Mostar […] The destruction of the Mostar bridge symbolizes the entire war of genocide being waged against the Muslims of Bosnia. This is a war against the harmonious coexistence of peoples.

Or another telling from a tourist guide (Lonely Planet, 2009):

> The beautiful old bridge (Mostar Bridge) […] was finished in 1566 and came to be appreciated as one of the world's engineering marvels. It survived the Italian occupation of WWII, but after standing for 427 years the bridge was destroyed in November 1993 by Bosnian Croat artillery. That was one of the most poignant, pointless and depressing moments of the whole Yugoslav civil war. (p. 132)

The presence of different versions of "the story of the Old Bridge" would force us "to choose among several interpretative options" (Derrida, 1980, p. 55). Considering the existence of comprehensive and contradictory discourse on the Old Bridge, this study will attempt to depict how those existing telling are indeed dialogic. The next section will focus on the first domain of dialogic narration which is the story itself.

3.1 The Intrinsic Dialogue

Authoritative tellings such as of UNESCO view the Old Bridge as a symbol of multiculturalism, arguing it depicts the coexistence of communities belonging to different cultural, religious and ethnic backgrounds. But considering the fact that the Old Bridge was destroyed by one of the ethnic groups for representing the Other, using the Old Bridge as a symbol of multiculturalism creates a paradox.

Indeed, the metaphor of the bridge reflects an inherent paradox, since it might alternatively be used as a boundary between the imagined worlds of self and the Other or as a connecting point between two different identities. Thus, narration of the Old Bridge represents a paradox since it might be interpreted both as a bridge dividing the east-Bosnian side from the west-Croat side and as a connecting bridge considering the intermarriages and the presence of different ethnic and religious groups on both sides of the bridge.

Throughout the Bosnian War, the city of Mostar had been a world of others, a world populated by residents who belong to different ethnic and religious groups. In Mostar, the boundary between self and other had been the rule where a Croat defined a Bosnian as other and vice versa. Accordingly, the Old Bridge as a public space had become the locus of the world of others. Throughout the Bosnian War, the society witnessed the killing thousands of people and the destruction of the Old Bridge for being a symbol of the other. The Old Bridge became a centre of military conflict partly due to what it meant for a specific group of people and what it represented for different groups of people. As Low (2000) asserts that "the position of the viewer–socially, politically, and physically–influences what can be and is seen" (p. 50). Accordingly, in Mostar, the cultural and ethnical positions of the viewers affected how they saw the Old Bridge. The Old Bridge became a subtext where the ghosts of history clashed for today's space.

The case of the Old Bridge also depicts Lefebvre's (1991) distinction between the representation of space (the conceived) and spaces of representation (the lived) as what it represented for Croats was quite different and in conflict with what it represented for Bosnians. Accordingly, throughout the Bosnian War, the Old Bridge transformed from a mental battleground into a physical battleground, due to the existence of two universes of discourse speaking with two voices. Then, the Old Bridge became a centre of military conflict partly due to what it means for a specific group of people and what it represents for different groups of people.

The existence of alternative narratives tends to deconstruct and even delegitimize the authoritative narrative. Yet, we cannot view those existing narratives in isolation, as they rather occur in reference to other narratives. Thus, dialogic narration implies a polyphonic discourse in line with the presence of multiple tellings of the Old Bridge. Accordingly, the residents of Mostar find themselves continuously experiencing a dialogic reality which is reflected in their questions regarding their past and future as well as present.

Shall we keep the Old Bridge as a symbol of multiculturalism or of cultural enmity? Fight or live peacefully? Live through preserving our ethnic boundaries without communicating with the Other? Or find ways to communicate and create harmony and coexistence?

Bruner & Gorfain (1984) emphasise the fact that "every story can be reinterpreted at a later date or retold in a different context, and new meanings will emerge" (p. 65). This assertion takes us to the following section which will focus on how narration of the Old Bridge evolved throughout history.

3.2 The Historical Dialogue

Looking in-depth to narratives of the Old Bridge depicts how the struggle between representatives of different ethnic groups as well as the international organisations over the meaning of the Old Bridge has evolved throughout time. In line with the recent conjecture, residents of Mostar–with the support of international foundations like UNESCO–rebuilt the Old Bridge to create a new identity and to change the negative perceptions of the city following the end of the war.

Following the Old Bridge's inclusion in the World Heritage List, its identification, protection and preservation as a cultural heritage site was accelerated. The process of reviving Mostar as a centre of global tourism started. In a sense, global economic challenges forced residents to redefine the image of the Old Bridge as a symbol of multiculturalism. The Old Bridge was transformed into a commodity space following its addition to the UNESCO (United Nations Educational, Scientific and Cultural Organisation) World Heritage List. Accordingly, the Old Bridge turned from the bridge dividing the Croatian side from the Bosnian side into a bridge connecting both sides. Cultural contestation was left aside while economic challenges, like attracting tourists and foreign investors, enabled to redefine the narrative of the Old Bridge.

In addition to UNESCO, another authoritative voice, the Tourism Association of the Federation of Bosnia and Herzegovina, emphasizes the multicultural character of the space and asserts that "even mentioning the bridge for years after it fell to the bottom of the Neretva River could invoke tears as it symbolized both the city and the country as a whole" (Tourism Association B-H, 2010).

Redefining the Old Bridge as a symbol of multiculturalism takes the narrative as objective truth; and further attempts to maintain the endurance of the meaning. In that sense, this version of the narrative tends to be monologic as it presents "one voice, one story, one truth" (Bruner & Gorfain, 1984, p. 67). Since any monologic narrative is free from challenging versions, alternative interpretations of the Old Bridge have emerged throughout history. Examining the Old Bridge sheds light on the hidden social and political memories of the place. Thus, the Old Bridge presents us with "a living history of local cultural meanings and intentions retained through a mnemonic process of environmental memory decoded in space" (Low, 2000, p.49)

As an illustration of challenging narrative, Grodach (2002) views the redefinition of the Old Bridge as "neo-colonization and creation of new market spaces through physical and economic intervention and privatisation" arguing that there was no reference to the Old Bridge as a symbol of multiculturalism until the destruction of the Bridge (p. 73). Grodach (2002) asserts that an Ottoman architecture was chosen because it was regarded as a convenient way to bolster the war-torn economy of the city despite the fact that the city is still ethnically divided.

Considering the existence of multiple interpretations and alternative narratives, the discourse about the Old Bridge changes over time and continues to evolve in line with historical, political and economic changes as well. As Bruner & Gorfain (1984) asserted that "the story responds to shifting cultural and historical contexts; and meanings reside not in the original but in present tellings" (p. 71). Narratives about a public space also reflect how an individual aligns himself with the space which will be examined next.

3.3 The Experiential Dialogue

The interpretation and meaning of the Old Bridge is attached to the material site that is the pedestrian bridge across the Neretva River. Stories on the Old Bridge not only imply historical facts but also create a geographical alignment. The material existence of the Old Bridge enables an individual to experience space as well as the imagination of space. For instance, when describing the Old Bridge, a newspaper article refers to people's activities on the bridge:

> Serbs and Turks, Croats and Jews, Greeks and Albanians, Austrians and Hungarians, Catholics, and Orthodox, Bogumils and Muslims–all had stopped at the same spot, rested on the same stone (Drakulic, 1993).

Although the Old Bridge is located on a given geographical point, dialogic narration is intended for mobility and extends far beyond the territory of the bridge. The Old Bridge attains meaning besides its context. As an illustration, when the Old Bridge was demolished, even individuals who had never been to Mostar before created a connection with the bridge.

The excerpt below denotes how the story of the Old Bridge was told through an individual:

> I was four when he [my father] wrote that postcard, and I know that he was certain that one day I would see and touch the Mostar bridge, too. My father was wrong. I did not make it. I foolishly thought the bridge would be there forever. So I never went to Mostar, never walked from one bank of the river to the other. The bridge that saw so many wars, survived so many years, no longer exists. It collapsed in a second on Nov. 9. All I have to remember it by are these three photos: before, during and after. (Drakulic, 1993).

The physical existence of the Old Bridge strengthens authoritative interpretations of the individuals assuming that the story itself is visible and touchable. Therefore, after the destruction of the Old Bridge, residents felt that they lost someone that they already knew, or something of their part. For instance, a girl named Melissa from Mostar describes how she felt after the Old Bridge was destroyed by pointing at the river:

> You cannot imagine what we felt like when the bridge was destroyed. We felt as if part of our soul had been destroyed. The Croats have a culture, a civilisation. How could they do this? This is also their culture [...] But there's something very strange. When we heard the bridge was gone, first we felt our soul had gone too, then we also felt much stronger (Fisk, 1994).

Another resident interprets the loss of the bridge as an attack to "contact between peoples" (Cohen, 1994); and some argued that "common identity and the hope of coexistence" were destroyed with the bridge (LaFranchi, 1994).

The narrative of the Old Bridge also turned into autobiography as its meaning is individualized. When the bridge was demolished, an individual asked: "knocking the Bridge down, did they manage to kill Mostar? Hopefully they didn't; but they did kill a part of myself" (Radovic, 1993).

A 70-year old resident of Mostar, Borjanka Santic describes his own experience of the Old Bridge as follows: "I enjoyed my first kiss on that bridge, I remember even now the stars and the moon shining down. I remember how we dropped stones into the clear water. Now this has all been wiped out" (Sudetic, 1993). Furthermore, some residents define the loss of the bridge as losing a close relative or losing a part of their being (Sudetic, 1993).

As the story of the Old Bridge is retold or the destruction of bridge is seen, individuals participate in the discourse. Throughout this process, the audience becomes an active participant and emotionally aligns itself with the space. For instance, the excerpt below illustrates how public space was experienced by an individual:

> Not so long ago, newspapers published photos of a massacre in the Bosnian Muslim village of Stupni Dol. One picture showed a middle-aged woman with a long, dark knife cut along her throat. I don't remember anyone crying over that photo, or others like it. And I ask myself: Why do I feel more pain looking at the image of the destroyed bridge than the image of the woman? Perhaps it is because I see my own mortality in the collapse of the bridge, not in the death of the woman. We expect people to die. We count on our lives ending. The destruction of a monument to

civilization is something else. The bridge, in all its beauty and grace, was built to outlive us; it was an attempt to grasp eternity. Because it was the product of both individual creativity and collective experience, it transcended our individual destiny. A dead woman is one of us; the bridge is all of us, forever (Drakulic, 1993).

The interpretations belonging to different individuals given above depict that dialogic narration prevails as people move through the space and/or imagine the space. Some of the individuals are residents who lived in Mostar and walked through the Old Bridge, some saw the Old Bridge on a postcard and imagined one day to cross the bridge, and some just watched the destruction of the bridge live on TV. Yet, the physical existence or non-existence of space plays a crucial role in their experience of the Old Bridge. This validates the argument that "stories, like ritual, are transformative insofar as they are experienced and performed" (Bruner & Gorfain, 1984, p. 73). Accordingly, narration of the Old Bridge is not only dialogic with itself and history, but also with the individuals who tell and experience the story.

4. Conclusion

The "bridge" metaphor has been a common norm in the self-perception of the Balkan nations, as they were located at the crossroads of civilizations (Grodach, 2002). The metaphor has been interchangeably used as a boundary between the imagined worlds of East and West and as a connecting point between Christianity and Islam. A 16th century pedestrian bridge, the Old Bridge, standing across the river Neretva and connecting both sides of the Mostar city in Bosnia and Herzegovina has been used as a narrative to (re)define Mostar throughout history.

This study focused on the concept of dialogic narration and depicted how narration of a specific public space, the Old Bridge, responds and interacts with itself, with communities and their histories, and with the self.

The study examined three stages of dialogic narration: intrinsically, by looking at multiple voices within the text; historically, through focusing on the changes occurred in the text in line with historical, political and economic changes; and experientially, by observing how individuals tell and interpret the text.

The case study of the Old Bridge enabled us to understand the complexities behind the reconstruction of a public space in the divided city of Mostar, where diverse groups attach different meanings to the same place. In a post-conflict society like Bosnia and Herzegovina, public space facilitates a central role in (re)defining the local identity and maintaining ethnic relations not only to preserve peace and stability in the country, but also to attract tourists and enhance the global identity of the city.

Public space not only exists physically (in material and architecture terms), but also imaginatively (in people's minds as an idea or as a symbol). Accordingly, the study depicted that narration of the Old Bridge is not static; it always changes, allowing multiple interpretations throughout time and depending on the context. As it is already asserted, "no story is 'a' story or 'the' story but rather a dialogic process of many historically situated particular telling" (Bruner & Gorfain, 1984, p. 57).

References

Andric, I. (1945). *The bridge over the Drina*. London: The Harvill Press.

Bakhtin, M. M. (1981). *The dialogic imagination: Four essays*. M. Holquqist, trans. C. Emerson, ed. Austin: University of Texas Press.

—. (1984). *Problems of Dostoevsky's poetics*. C. Emerson, trans. and ed. Minneapolis: University of Minnesota Press.

—. (1986). *Speech genres and other late essays*. C. Emerson and M. Holquist, eds. V. W. McGee, trans. Austin: University of Texas Press.

Bourdieu, P. (1977). *Outline of a theory of practice*. Cambridge, MA: Cambridge University Press.

Bruner, E. M., & Gorfain, P. (1984). Dialogic narration and the paradoxes of Masada. In E. M. Bruner (Ed.). *Text, play and story* (pp. 56-79). Prospect Heights, IL: Waveland Press, Inc.

Bruner, J. (1990). *Acts of meaning*. Cambridge, MA: Harvard University Press.

Carmona, M. et al. (2008). *Public Space: The Management Dimension*. London & N.Y.: Routledge.

Carr, S., Francis, M., Rivlin, L. G., & Stone, A. M. (1992). *Public Space*. New York: Cambridge University Press.

de Certau, M. (1984). *The Practice of Everyday Life*. Berkeley: University of California Press.

Derrida, J. (1980). The law of genre. *Critical Inquiry*, *7*, 55-81.

Donia, R. J., & Fine, J. V. A. Jr. (1994). *Bosnia and Hercegovina: A tradition betrayed*. London: Hurst & Company.

Drakulic, S. (1993, Dec. 24). Bosnia-Herzegovina: A bridge too dear. *The Globe and Mail.*
Dube, N. (1993, Nov. 19). Letters to the editor: Barbarism reigns. *The Daily Youmuri.* p. 6.
Duncan, J. S. (1990). *The city as text: The politics of landscape interpretation in the Kanyan Kingdom.* Cambridge: Cambridge University Press.
Federation of Bosnia Herzegovina Office of Statistics (2009). *Population of Mostar*. Retrieved from http://www.fzs.ba/Dem/Popis/Nacionalnost%20opcine%20Popis%201991.pdf
Fisk, R. (1994, June 19). Follwing the devil's track. *The Independent.* The Sunday Review, p. 4.
Foucault, M. (1975). *Discipline and punish.* New York, NY: Random House.
—. (1984). Of Other Spaces. *Architecture/Mouvement/Continuite.*
Grodach, C. (2002). Reconstituting identity and history in post-war Mostar, Bosnia-Herzegovina. *City*, *6* (1), 61-82.
Gupta, A., & Ferguson, J. (1992). Beyond "Culture": Space, Identity, and the Politics of Difference. *Cultural Anthropology*, *7* (1), 6-23.
Harvey, D. (1985). *Consciousness and the urban experience.* Baltimore, MD: John Hopkins University Press.
—. (2006). The political economy of public space. In S. M. Low & N. Smith (Eds.). *Politics of Public Space* (pp. 17-34). London: Routledge.
Katriel, T. (1994). Sites of memory: Discourses of the past in Israeli pioneering settlement museums. *Quarterly Journal of Speech, 80* (1), (1-20).
Kotler, P., & Gertner, D. (2002). Country as brand, product, and beyond: A place marketing and brand management perspective. *Journal of Brand Management*, *9*, 249-261.
LeFebvre, H. (1997). *The Production of Space.* New York: Blackwell.
Lonely Planet (2009). *Western Balkans.*
Lotman, Y. M. (1988). Text within a text. *Soviet Psychology, 26* (3) (spring), 32-51.
Lotman, Y. M. (1990). *Universe of the mind: A Semiotic theory of culture.* Bloomington: Indiana University Press.
Low, S. M. (1996). Spatializing culture: The social production and social construction of public space in Costa Rica. *American Ethnologist*, 23 (4) (November, 1996), 861-879.
—. (2000). *On the plaza: The politics of public space and culture.* Austin: University of Texas Press.

Makas, E. G. (2007). *Representing competing identities: Building and rebuilding in post-war Mostar.* (Unpublished doctoral dissertation). Cornell University, New York.

Nielsen, G. M. (1987). Reading the Quebec imaginary: Marcel Rioux and dialogical form. *The Canadian Journal of Sociology / Cahiers canadiens de sociologie, 12* (1/2) (Spring), 134-149.

Pašić, A. (2006). *The old bridge in Mostar.* Gračanica, Bosnia and Herzegovina: Grin.

Radovic, D. (1993, Nov. 15). A bridge too far. *The Age.* Opinion, p. 14.

Rapoport, A. (1982). *The meaning of built environment: A nonverbal communication approach.* Arizona: The University of Arizona Press.

Schuman, M. A. (2004). *Nations in transition: Bosnia and Herzegovina.* New York, NY: Facts on File, Inc.

Sudetic, C. (1993, Nov. 10). Croatian artillery fire destroys 16-th century Ottoman bridge. *The Houston Chronicle.* p.22.

Tourism Association Bosnia and Herzegovina. (2010). *Mostar.* Retrieved from http://www.bhtourism.ba/eng/mostar.wbsp

Tunbridge, J. E., & Ashworth, G. J. (1996). *Dissonant heritage: The management of the past as a resource in conflict.* Chichester: Wiley.

UNESCO. (2010). *World Heritage List.* Retrieved from http://whc.unesco.org/en/list/946H

Wertsch, J. W. (2000). Narratives as cultural tools in sociocultural analysis: Official history in Soviet and Post-Soviet Russia. *Ethos*, *28* (4), 511-533.

White, H. (1987). *The content of the form: Narrative discourse and historical representation.* Baltimore: Johns Hopkins University Press.

CHAPTER XVI

IMAGINE MILAN: AN AUDIOVISUAL DESIGN THINKING APPROACH TO THE IMAGE OF THE CITY

MARIANA CIANCIA AND WALTER MATTANA

The great developments in social, economic and urban planning fields are deeply changing the identity of the city, in Milan as well as in other metropolises. A complex system in which physical emergencies and perception of the image of the city converge (Lynch, 1960): what we should call a *collective mental synthesis* (Bonini Lessing, 2010).

Historically, the urban field has always been an area of interest common to many studies–architecture, city planning, sociology, semiotic, economics, place marketing–and in the last decades communication has also been involved in this research. These studies highlighted how socio-cultural development has changed the perception of the urban area, with the subsequent creation of an image of the city built through a continuous overlap of sensory levels: between real life experience and media polyphony (created by analogue and digital artefacts). A changing city where information exchange feeds the communication flows necessary to maintain the identity of a more and more plural society, and where developments leaded by new technologies affect consumption, experience and social habits (Bonini Lessing, 2010).

As a consequence, the great challenge of the present urban fabric arises, which is related to the promotion of coexistence within a plural society through a new approach: a society that needs planning, involvement and collaboration like the Anglo-American governance (Innerarity, 2008).

Therefore, if society demands behaviours able to match up with new cultural and politic pluralism, and starting from the assumption that the city is closely related to citizens who are as important as the physical framework, we need a strategy to tackle the *deep diversity* (Taylor, 1992).

The aim of this work is to suggest a design approach to these emergencies in order to overcome the problems in understanding and foster social dialogue, using audiovisual tools.

1. Imagine Milan: a communication system for the social dialogue

Imagine Milan is an educational and research project started in 2009, which involves professors, researchers and students of the School of Design – Politecnico di Milano. The aim of this research is to delineate a project able to experience audiovisual formats and a communication system capable of promoting social dialogue and innovation.

The first phase is related to the development of audiovisual communication artefacts for the territory's listening, which allowed us to map 20 districts in Milan.

On the first year (*Walking in Milan*, 2009-2010) we worked on *Bovisa*, *Isola*, *Sempione*, *Brera*, *Lazzaretto*, *Città Studi*, *Navigli*, *S. Ambrogio*, *Quadrilatero della Moda*, focusing on the theme of sustainable mobility in collaboration with *Comune di Milano* (Milan Municipality).

The second year (*Street Tales*, 2010-2011) was not related to public administration and referred to the quality of life in Milan, aiming to activate relationships with stakeholders in the urban fabric: *Garibaldi*, *Bicocca*, *Via Farini*, *Ortica*, *Ticinese*, *Dergano*, *Via Padova*, *Via Sarpi*, *Corso Buenos Aires*, *Tortona*.

On the third year (2011-2012) we worked again on the audiovisual output; however we proposed a different approach starting from themes related to city transformation: Smart City, Green City, Sustainable Mobility, Multiculturalism, Social Memory, Suburbs.

1.1 Audiovisual Communication Formats

The *Imagine Milan* research intends to test audiovisual and co-design practice for social innovation, focusing on the potentialities of audiovisual storytelling within social communication through three different narrative formats. We can highlight the main project phases (Figure 1).

The first phase–*Listening*–has the aim of exploring the development territory. Therefore designers should listen to the area through the documentation of case studies and best practices; video interviews with citizens, workers, associations, craftsmen and companies; the search of historically and contemporary iconographic repertories, useful to reconstruct the memory and social imagery of the place.

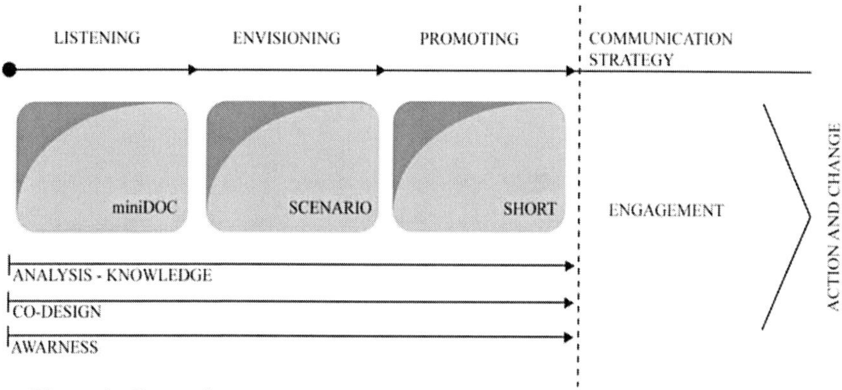

Figure 1 - Research process

The output of this phase is the *miniDOC* (Figure 2). An audiovisual format able to tell in a short length of time (five minutes) the most important aspects emerging from the previous research and analysis work. A basic exploration able to build not only audiovisual artefacts, but also design processes. A crucial point that allows the designer to gather and organise the key elements in narrative and aesthetic clusters: hence the investigated field becomes a source of inspiration for the development of the work.

Figure 2 - MiniDOC

The second phase–*Envisioning*–has the aim of providing an epistemological and aesthetic contribution to envisioning a sustainable future, towards an *Audiovisual Design Thinking* approach (Piredda, 2008).

Design Thinking is a style of thinking that allows designers to combine empathy and rationality in the context analysis and to use creativity in concept generation. In fact, according to Brown (2008, p. 88), it is possible to describe the design process "metaphorically as a system of spaces rather than a predefined series of orderly steps. The spaces demarcate different sorts of related activities that together form the continuum of innovation".

For this research we start from the assumption that the audiovisual approach is capable to promote a dialogue among people, relying on the storytelling ability to foster different perspectives. An approach akin to *Audiovisual Thinking: A Journal of Academic Videos*, an online journal whose aim is the experimentation in new forms of academic contributions (online www.audiovisualthinking.org).

The output is the *Scenario* (Figure 3): a three-minute video based on a narrative structure in three acts. A short audiovisual format whose aim is the visualization of abstract concepts for activating negotiation tables and conversations among stakeholders.

Figure 3 - Scenario

The third phase–*Promoting*–intends to promote a sustainable city life and its values and benefits through a typical advertising output (Figure 4): an audiovisual *short* format (thirty seconds).

Figure 4 - Short

The production of each output is characterized by the use of a typical brand communication tool: Copy Strategy. A document that delineates a communication strategy through some points like: communication concept (Promise); explanation of the promise (Reason Why); motivations able to support the promise (Supporting Evidence); description of the Target; Creative Concept; look and feel (Tone of Voice). Our aim is to push students to understand how each output becomes an integral part of a contemporary communication process: the reason why they have to explain every planning choice.

In general, the first research phase tried to interweave the communication project with the experimentation of co-design practices. In fact, this step ended with a workshop at the *Urban Centre* in Milan in order to experiment the use of audiovisual artefacts to engage stakeholders for envisioning a sustainable future.

1.2 Transmedia Communication System

After the development of audiovisual formats, which allowed us to map 20 districts in Milan, our research group is working on a participative communication system where to file these outputs.

Nowadays we live in a society enriched by a growing mass of information spread by multichannel and multimodal devices. A complex scenario in which it is important to study the evolution of the existing communication systems, in order to understand the new processes, and how media and languages have broken their historical isolation.

In contemporary society it is possible to highlight the developments in the communication field that are affecting society by exploring three main areas: audiences, technologies and user engagement.

The first change, taking place in a script society (Toschi, 2011), is related to the audience using communication systems. During the last decades we have witnessed developments that have given a new relevance to the storytelling practice. Compared to the last decades, the contemporary strength is the rising importance of new narrative structures with stories able to provide not only a linear path–with a beginning, middle and end–but also multiple points-of-entry.

The second change concerns the technological field: a development that leads us to interact with a digital vision of reality. Nowadays the remarkable acceleration in technological development has provided unimaginable instruments to the communication strategy. Hence, a designer needs to understand how to make a proper use of these tools in order to develop communication systems able to combine different media (on-line/off-line) to engage audiences.

The third change is related to participatory processes. In the last decades we have been witnessing a shift from a *platform-centric vision* (Bernardo, 2011), to a *user-centred vision*, but now it is necessary to push the development toward a *community-centred vision* to allow the growth of collective consumption and community engagement. Moving the focus from technology to audience experience, it is possible to find a balance between the two engagement degrees: as a consequence, people can enjoy the best contents spread by the most suitable devices. For these reasons, the designer has to learn how to manage the new communication processes: shifting his position from an executive role to a supporting role, while the generative process increases. This is because the audiences are becoming aware of their lead role in contemporary communication activities.

Three fields–audience, technology, engagement–that are living in a situation of interchange within the age of convergence (Jenkins, 2006) and whose development has given rise to conceptual spaces that affect the *mediascape*: *Storytelling Space*, *Participative Space* and *Experiential Space*.

Storytelling Space and Participative Space refer to new narrative structures that give people more tools to tell and to listen to stories. A contemporary reality where audiences and technological transformation create an Experiential Space able to envision transformation, translate cultural repertories and produce a brand new and shared vision: a collaborative network powered by stakeholders (Lévy, 1997).

Our answer to these emergencies from a design point of view is a *transmedia* approach. According to Jenkins (2011), who in further reflections on the transmedia term (used by itself) defines it "one way of talking about convergence as a set of cultural practices", in this research it is proposed as a shift in the use of the word from an adjective to a noun. The idea is that there is a *Transmedia Practice* (Dena, 2009) that allows audiences not only to access contents in a different way (Murray, 1997), but also to participate in the meaning-making process, with a subsequent change in the relationship between the mainstream media (*top-down*) and participatory culture (*bottom-up* or *grassroots*).

2. Audiovisual storytelling for social conversation

Cities live into a two-fold dimension. The first one is the physical dimension of architecture and city planning, which materially rules and organises territories and spaces. The second one is the immateriality of its imagery, which is built and consolidated over time through a complex system of representation made by narrations, images and memories. In this sense, cinema and the other forms of audiovisual storytelling are essential instruments for the interpretation and comprehension of the urban phenomenon (Galbiati, 1989).

Just like any other form of narrative expression, cinematographic storytelling does not record data of reality, but necessarily tells stories through a *mise en scène*. To the objectivity of scientific description audiovisual language it opposes narrative subjectivity, which blends real and imaginary elements in order to give a representation of reality that is not exact, but authentic–and therefore likely and plausible–as we perceive it. The two extremes are not in collision, but are integrated as Bruner (1985) theorized about human cognitive processes.

Narrative thinking filters reality through the consciousness of its characters in order to play and conceptualize human experience. Similarly, audiovisual storytelling is configured as a system for the organisation of thoughts and knowledge, then it performs a number of key functions at both individual and community level. We can summarize these key functions as follows:

- *Frame*: all stories provide a *frame of reference* that includes the narrated events and in which the spectator can place its own life experiences, making comparisons. This function is even more strengthened in cinema and audiovisual tales, in which the concept of *frame* (related to the idea of *a window on reality*) takes a real *physical* identity that isolates and highlights the value of images (Elsaesser & Hagener, 2007).
- *Memory and identity*: audiovisual tales represent the storage of collective, individual and autobiographical memories through voices, words, images, gestures and languages. Those shared memories are going to become a collective identity.
- *Conversation*: storytelling as a tool for sharing and dialogue between individuals and organisations, that facilitates and guides social behaviours in terms of responsibility, integration, best

practices, sharing of knowledge and values towards change.
- *Imagery*: audiovisual storytelling practice makes society able to review its past, codify and frame the present and represent its future expectations.

Audiovisual storytelling enables all these functions from an essential element: dramatization. Each story is actually a representation of a conflict. Characters are individuals in a perpetual struggle with their fate and the environment around them, suffering and struggling to achieve a specific goal.

Drama is the real drive of any narrative, the element of the discourse that creates the bond with the audience, allowing them to portray themselves and their own experience in the history that they are listening to.

The mechanism of drama not only applies to the narrative genres of fiction or entertainment (film, TV series, reality shows), but it is also functional in the context of social communication and management referring to citizens and stakeholders. Representing the transformations taking place, promoting innovative services or spreading new behaviours, means first of all clashing with the resistance, fears and prejudices that the public opinion shows against any process of change.

Nowadays, communication cannot simply reassure the audience or enhance the performance of objects and services. In fact, storytelling has to tell this emotional conflict and then solve it, triggering a process of identification that turns the story-listener into an active subject within the communication process.

The *Imagine Milan* project tried to test and verify the potentialities of audiovisual storytelling within social communication. In particular, miniDOC and scenarios clearly show that the identity of an area is built right from the representation of personal and collective histories of its inhabitants. A set of polyphonic images, faces, voices, gestures and characters in a continuous comparison (sometimes, indeed, stressful and conflictive) among the historical memory, the complexity of the present and the future expectations.

3. Conclusions

In these years we have tested and validated the use of audiovisual language, historically used as a knowledge instrument. We recognized its dual status:

- *A tool able to read the present* – Nowadays the citizens feel the need for instruments capable to give them whole *mnemonic* images, which are lost due to the growing mobility and isolation phenomenon;
- *A tool able to suggest a dialogue on a possible future* – The audiovisual artefact could aid to create a new collective imagery related to sustainability: a kind of *pedagogy glance* that will make people aware and involved in the city's daily life.

An Audiovisual Design Thinking approach able to document physical, social and cultural reality, and also to propose additional values, through which we can promote transformation visions: quoting Rossi Doria, a *value-added interpretation* (Leone, 2010).

The next phase, which involves the integration of audiovisual artefacts in a participatory communication system, started with the identification of a possible answer able to transform the contemporary criticality into strength points: the designing of a *transmedia communication system*.

Communication designers need to understand how to develop this project starting from three crucial points:

- Transmedia storytelling–that is the flow of content across media (Jenkins, 2003; Giovagnoli, 2009, 2011; Davidson, 2010; Perrone, 2012);
- Involvement of different media (on-line, off-line, new and social media)
- Audience Engagement–in order to foster the creation of collaborative network powered by stakeholders (Lévy, 1997).

References

Bernardo, N. (2011). *The producer's guide to Transmedia. How to develop, fund, produce and distribute compelling stories across multiple platforms*. Lisbon-London-Dublin-Sao Paulo: Beactive Books.

Bonini Lessing, E. (2010). *Interfacce Metropolitane*. Milano: et al./edizioni.
Brown, T. (2008). Design Thinking. *Harvard Business Review*, *86* (6), 84-92. Retrieved from http://hbr.org/2008/06/design-thinking/ar/1
Bruner, J. (1985). *Actual Minds, Possible Worlds*. Cambridge: Harvard University Press.
Davidson, D. (2010). *Cross-Media Communications: an Introduction to the Art of Creating Integrated Media Experiences*. ETC Press.
Dena, C. (2009). *Transmedia Practice: Theorizing the Practice of Expressing a Fictional World Across Distinct Media and Environments.* PhD dissertation. University of Sydney, Sydney, Australia.
Elsaesser, T., & Hagener, M. (2007). *Filmtheorie. Zur Einfürung*. Hamburg: Junius Verlag GmbH. (it. tr. *Teoria del film. Un'introduzione*, Einaudi, Torino, 2009).
Galbiati, M. (Ed.). (1989). *Proiezioni urbane: la realtà dell'immaginario*. Milano: Tranchida.
Giovagnoli, M. (2009). *Cross-Media. Le nuove narrazioni*. Milano: Apogeo.
—. (2011). *Transmedia Storytelling. Imagery, shape and techniques*. ETC Press.
Innerarity, D. (2008). *Il nuovo spazio pubblico*. Roma: Meltemi.
Jenkins, H. (2003). Transmedia Storytelling. *Technology Review*. Retrieved from http://www.technologyreview.com/biomedicine/13052/
—. (2006). *Convergence Culture*. New York: New York University Press.
—. (2011). "Transmedia 202: Further Reflections", Confessions of an Aca-Fan. The Official Weblog of Henry Jenkins, August 11, 2011, http://henryjenkins.org/2011/08/defining_transmedia_further_re.html
Leone, D. (2010). *Sequenze di città. Gli audiovisivi come strumento di studio e interpretazione della città*. Milano: Franco Angeli.
Lévy, P. (1997). *Collective Intelligence. Mankind's emerging world in cyberspace*. New York: Basic Books.
Lynch, K. (1960). *The Image of the city*. Boston: MIT Press.
Murray, J. (1997). *Hamlet on the Holodeck: The Future of Narrative in Cyberspace*. Boston: MIT Press.
Perrone, D. (2012). *What Jeff has not said*, "Subvertising. La sovversione della pubblicità" (year VI, number 50). Retrieved from http://subvertising.it/
Piredda, F. (2008). *Design della comunicazione audiovisiva. Un approccio strategico per la "televisione debole"*. Milano: Franco Angeli.
Taylor, C. (1992). *Multiculturalism and the Politics of Recognition*. Princeton: Princeton University Press.
Toschi, L. (2011). *La comunicazione generativa*. Milano: Apogeo.

CHAPTER XVII

DANTEAN SPACE IN THE CITIES OF CINEMA

AMEDEO D'ADAMO

We... nursed within ourselves sullen fumes,
and (thus) come to misery in this black ooze.
(Dante's Inferno, Canto 7, line 123).

Louis Kahn, the seminal master Architect, once famously held up a brick and asked his students "What does a brick want to be?" It was a pure Kahn teaching moment; his gnomic question captured something of architecture's phenomenological pressures–the demands of materials, the connection between shape and structure, even the hidden quality of the "meaning" of great architecture–while at the same time it hinted at the source of his own primal monumentalism. It seemed to indicate that somehow architecture built itself, or could do so, without concern for the lives and desires of the humans who live in it; a kind of Architecture without Architects or maybe even human beings.

Kahn's question is a rich and provocative one for Architects, but the question is fundamentally the wrong one for anyone who builds a city for the movie business. If Kahn were building a set for a narrative film and asked "what does this brick want to be?" the smart Production Designer or Producer or Director or Screenwriter would reply: "Well, hold on, what does the protagonist or antagonist want that brick to be?"

This central fact helps to explain why the cities of cinema are such strange places. We all can sense that when cities are cast in a film, they change; they somehow stop being themselves. It doesn't matter if they are actually Potemkinesque constructions–crafted on a backlot or a hard-drive like *West Side Story* or *Casablanca*–or if they pulse with real and authentic life as if almost accidentally discovered by the camera, as in *Rome Open City*, *Salaam Bombay* or the first act of *Slumdog Millionaire*;

once they are harnessed to a protagonist's story, somehow all cities transition into a strange new phantasmagorical version of themselves, a phantasmagorical environ for the story's characters.

But what more can we say about this strange change? Quite a bit. It turns out, of course, that there is a history to the invention of architecture in literature and drama, a history that is tied up with the inventions of Dramatic theory and of Interior space itself.

In fact there are actually three distinct layers of entanglement, each with its own specific Dramatic architecture. Sometimes a character is just within a city, while sometimes a character is struggling with or for a city, and sometimes a character while within a city is also in a dramatic sense also within him/herself, in that the city has become a kind of projection of the character's dramatic inner space.

To understand this more deeply, we will first detail the Aristotelian principles underlying these kinds of cities. Then we will explore all three versions of the cities of cinema, all three versions of dramatic space. Then we will focus on the third kind, that most entangled, intense form of dramatic space that we will call Dantean Space, a category of aesthetic space largely invented by Dante when he architected the three zones in his *Divina Commedia*. That is the plan.

However, before we start I think it will be helpful to add a short note about my methodology here. Until the paper's conclusion I stay away from even the simplest sociological concepts such as Identity and Subjectivity, instead limiting the investigation to a tight diet of dramatic screenwriting terms such as Character. The intent is to begin with a stripped-down NeoAristotelian methodology, that of the story phenomena we are trying to explicate, and then use that to reveal the logic, practice and impact of three forms of story Space, and then ask about the implications of these forms of Space. I hope my overall logic is clear; Aristotle's story theory is deeply important for both Dante and for the majority of today's narrative filmmakers and, as a rule, when you set out to theorize about any field or activity, it is always best to initially stay as much as possible within the actual pragmatic working vocabulary of that field instead of immediately importing concepts from another body of work. In this way you are more confident of capturing the essence of whatever the ends of that field's activities are. However, I don't think Aristotle is the horizon of useful thinking on aesthetics and culture; once we have tried to deconstruct the phenomenon in this way, I then argue that the results have a real relevance for the broad and urgent project of Cultural Studies.

1. Taxonomy of Aristotelian story elements

And so to trace out the ontological nature of cities in stories, we first look to the philosopher who Dante himself ranks as the first among philosophers: Aristotle.

Let's begin by detailing the four core Aristotelian possibilities in a narrative; this will inform us about the four ways a city in a film (or really any dramatic element at all) can function in a Character's Arc. In this compendium we are using a city as an example of a story element, though of course any dramatic object or person such as a gun, a car, an ex-boyfriend or a gorilla can also serve as a story element.

In Aristotelian dramatic theory, a city can serve the drama in any of the four following ways. The first two are often known as "Outer" because they generally involve the Plot, while the second two are often labelled "Inner" because they grow from issues of the character and are grounded in the protagonist's desires, secrets, fears, dreams and psychological history.

A city can serve:

- *As an outer antagonist.* This by itself is the most prevalent form of the Dramatic City, found in films where the city is a literal concrete problem for the Protagonist. This problem can be of many kinds: the city can for example be the threat itself that must be dealt with–think of the Death Star in *Star Wars*, for example–or the city can be a Maze. In The Maze model, the core idea of prison movies (such as *A Man Escaped* or *The Green Mile*) is expanded to fit an entire city, as we see in a recent example such as *The Adjustment Bureau* but also in older ones such as *Escape From New York*. In these movies the protagonist must, in a very concrete way, fight to triumph over the authorities, structures, buildings, streets and infrastructure of the city itself. The thriller *Eagle Eye* is also of this sort, as is *The Matrix*, where the city is a literal manifestation of the hegemonic robot (i.e., white-) culture that our hero is trying to destroy.

- *As an outer goal.* This is the second kind of Dramatic City, and here the city's approval or realization represents the protagonist's concrete, physicalized goal. There are many examples of this; for example the city may be under threat and must be saved. In such cases, if the protagonist can vanquish the bad guys, he or she will often ascend to a literal or metaphorical throne and gain great recognition–recently we saw this model in *The King's Speech* but also in the animated film *Rango*. A variant of this is the Plague city, a city that is under The Curse To Be Lifted by the Protagonist, who wants to become the saviour and hero of the city by ending this plague; the 2007 film *I Am Legend* is of this sort (though in this way it is a change from the original novel.) More rarely the city can be a goal of the Protagonist who is its architect; it is an expression of the best of the protagonist, as in *Bugsy*.

Hopefully, by first focusing on these Plot-oriented "Outer" categories we have now illustrated how a city can serve as a Dramatic Space. There are, however, two other narrative possibilities that we now want to focus on, because these two character-based roles lie at the very core of that extreme form of Dramatic space we are calling *Dantean* Space.

A City can also play a dramatic role:

- *As an inner antagonist.* In such cases it represents the Antagonist side of the soul-problem, where it somehow manifests an inner problem. Usually this problem is a fear, a lack, a shameful or dangerous secret that must be suppressed, or a dark desire that the protagonist is wrestling to overcome in himself. *The Graduate* is clearly struggling with L.A., as does the character of Marcello with Rome in *La Dolce Vita*. Three other such cities–all clearly Dantean in their conception–are the settings of David Fincher's films *7*, *The Game* and *Fight Club*. We can also see this realized in a film like *Dark City*, and in many interesting ways in a film like *Inception*.

- *As an inner goal.* Here the city represents the protagonist side of the soul-problem. It embodies the hope, dream or desire of the protagonist. Think for example of the spiritual allure that the heady artistic Paris of the 1920's has for the protagonist of Woody Allen's *Midnight in Paris*; the city is alluring precisely because it embodies in concrete fashion a social and aesthetic vision of a vivid artistic life, the dream, ideal and hope of the main protagonist. That city is contrasted vividly against the present-day cities of Paris and Hollywood where he now lives with his conservative fiancé, in order to dramatise what is very much missing from his current life. His conflict is essentially a conflict over the kind of city that he belongs to and that he should commit to, and the organic end to his story will be marked by his commitment to the right city, the one where he can finally be true to himself.

Note that our taxonomy covers all story genres. Certain genres are prone to take advantage of certain dimensions, but there is no singular one-to-one correspondence; these are really instead the deep options of any Aristotelian narrative. Also, like any story element, a city can function dramatically on both an External and an Internal level at the same time.

2. The gamut of dramatic architecture: DisPassionate space, Dramatic space and Dantean space

Let's look at each kind of dramatic architecture in turn. The gamut ranges from the least-involved relationship, a kind of space we might call DisPassionate space, where none of these four roles is being played by the city itself; the city is not important to the plot or to the character's inner goals. Then there is another kind of city that possesses something we can label Dramatized Space, which uses the first two of our categories–here the city is functioning as either an Outer Antagonist or Outer Goal of the protagonist of the story. And then there is a third kind of city, its most extreme and dramatically intense form, which manifests the character's Inner Antagonists and Inner Goals, the last two categories of our taxonomy. We are now describing the phenomenon we will call Dantean Space.

Let us illustrate these three kinds of cinema cities with a few specific examples.

In DisPassionate cities, the Spaces have little to no role in the drama, expressing neither the plot nor the personal struggles of the protagonist, and so they serve largely as a backdrop for the story they contain. We call these phantasmagorical cities DisPassionate because, even in their new screened life, they still manage to retain a kind of indifference or lack of entanglement; they still stand apart from the hero's struggle. Think of the cities of neorealism, of *Los Olvidados* or the New York City of *Midnight Cowboy* or *Taxi Driver*, or think of the neutral ambience of Barcelona in *Vicky Cristina Barcelona*. However spectacular they are on screen, these cities retain a kind of dramatic neutrality in the story, defined by the way they stand apart from the characters' desperate struggles, yearnings, hopes and point of view.

This can be contrasted with Dramatic cities, which clearly use the city as either an Outer Antagonist or an Outer Goal of the protagonist, but where the protagonist's inner struggle is playing out in a different arena. London in *The Kings' Speech* is a good example: the fate of the city (standing in as it does for all of Britain) hinges on the question of whether the King can perform as a King. The city is thus crucially represented in its role in the plot–it is what Hitler threatens. But our King's real inner conflict is not with the city itself; his emotional struggles all play out in the conflicts with his father, his brother, his friend the speech therapist, and of course at core with his efforts to overcome his childhood memories of abuse and his resulting lack of confidence and inability to speak clearly.

In contrast to these two kinds of cities, there is a third form of Space, a kind of extreme dramatic form that we are calling Dantean Space, where the city embodies the character's Inner Antagonist or Inner Goal, the third and fourth categories in our taxonomy. This rather abstract-seeming fact has a remarkable result; in these Dantean cities, Space itself becomes an extension, an expression and a physical projection of the character in it. Think of *Amélie*'s charming and peculiar Paris, filled as it is with the magical logic of her own imagination and yearnings, geographically and psychologically entangled as it is in her memories, projects and problems. Or think of Ripley's dripping, organic-yet-poisonous space colony in *Aliens*, where the alien creatures and their lairs are vicious living emblems of Ripley's deepest fears about sex, motherhood and human bonds. Each of these women's struggles with geography is a struggle with *themselves*.

A moment's reflection will reveal the difference in kind between these spaces and those of either *Vicky Cristina Barcelona* or *The King's Speech*. The city of Dantean Space serves not just as the host or backdrop for its hero as it might if it were a DisPassionate city, nor is it simply a Dramatic City that is simply richly realized in its Production values or "tone" (though it usually is). Rather, a Dantean city lives vividly and dramatically as the living spectacle of the soul of its protagonist. Such cities are not simply mirrors of the protagonist's fears or hopes, nor are they simply usually more spectacular than their neutral, indifferent sister-cities. Instead, these cities are spectacular in a different, deeper sense because in some very specific sense they are places that have been invested with a

Romantic energy so that they now intensify our cinematic experience in a very specific way–their spectacle actually takes us closer to the dramatic Point Of View of the film's protagonist.

Such cities are intense cinematic experiences, the kind you do not forget. And for only this reason it would be worthwhile to try to understand their peculiar dramatic architecture. But because of this very vividness of experience, they also become important in an economic sense. It is because they are both such intense cinematic experiences *and* have real-world references in their original, "native" city that they can activate whole sectors of the Cinema Tourism economy. When positive, as in *Amélie*, they produce hordes of cinema tourists, as that film did for the Parisian neighbourhood of Montmartre; viewers find the experience so rich that they spend vast amounts to travel to the "original" location in an effort to finally live the phantasmagoria. And even when negative, these Dantean Spaces can have concrete real-world effects, as for instance the darkly romantic, moody film *The Third Man* which helped make Vienna an international tourist destination.

For all of these reasons it is worth investigating both the ontological history and the dramatic architecture of these cinematic cities. They are at core best understood as a peculiar kind of dramatic device, one often used in the Romantic tradition and, I will argue, we owe their existence to the remarkable reinvention of Aristotelian story-craft by Dante in his *Divina Commedia*. The peculiar and intense horror of Ripley in her Alien-infested tunnels owes something to the tragic situation of Francesca and Paolo, lovers who are forever condemned to be blown about together but unable to touch, to the spectre of Ugolino gnawing the still-sentient skull of his tormentor Ruggieri, and to so many other unwilling residents suffering in Dante's city of Dis. The hordes of protagonists such as Liz Gilbert of *Eat Pray Love* are all the offspring of Dante's trip through Purgatory, and *Amélie*'s lovely version of Paris has its own debt to Dante's grand Space of altering love that is his Paradiso.

That, at least, is my claim. To see this more clearly, let's look at the particular invention that Dante makes, the innovation that lies at the core of so many of his powerful architectural-dramatic constructions. For he is the great inventor, the Ur-Architect who created phantasmagorical Spaces where the protagonist's Inner Antagonist and Inner Goals find a physical and dramatic expression in the architecture and spaces.

3. Dantean space

Let us first detail Dante's multi-layered creation of a Judicial Architecture, the overarching idea that also delineates Hell from Purgatory and both spaces from Heaven, the concept that defines the three grand spaces of the Commedia. It is simply this: in this Architecture, your life story not only determines your punishment–that is true for many in Hell–but at Dante's best moments your life story in fact becomes *physically manifest in a kind of self-created environment.*

Of course Dante was not the first one to conceptualize Hell itself. Homer, for instance, offered a comparatively shallow dramatic version. In the Odyssey, Hell is a very simple DisPassionate Space; Odysseus finds only an empty, alienating, dramatically inert plane where the wraithlike souls are condemned to drift about. Moreover, here it doesn't really matter who you are or what your life's drama was, whether you led a virtuous life or not; everyone who dies is just dumped here without discrimination and without possibilities to make a future story for themselves. There is no entanglement between their past lives and this present space–everyone is in the same boat in a place where nothing happens. Using our Aristotelian categories, Homer's Hell is a DisPassionate space, and perhaps it is not surprising that, architecturally-speaking, it barely exists.

Dante's Hell by comparison is an astonishing and complex creation, an architecturally–and dramatically–elaborated city. On its grossest physical level, Hell's layers of grand descending rings reflect a kind of marriage between City Planning and a vast penal code, with the zoning of large areas defined by the sins they hold. But within these rings many shades of subtlety appear; though in some areas people are grouped by their crimes and then punished simply in a kind of pecuniary or pre-legislated torture, there are many singular, sublimely dramatic exceptions. Often a soul here is shunted into a space or situation or structure that is a very specific mirror of the sins and guilty acts he or she committed in life, and it is no coincidence that these exceptions are the great dramatic figures of Dante's story.

Consider for example the case of Ugolino, a man who in life was a traitor and who was himself betrayed by the evil Ruggieri. Ruggieri then locked him in a tower with his children and starved them all, so that they died and Ugolino was finally forced to eat their corpses. As a direct result of this biography, these two traitors are now found together locked in ice up to their necks, and Ugolino is gnawing angrily at the skull of Ruggieri. We understand that both of these men are the architects of this horrifying situation, but we also are to understand that the paralyzing, unmovable nature of Ugolino's fury–a very understandable emotion in the circumstances–is physically manifested by the encasing ice. The frozen ice is itself both part of his crime and part of his punishment.

The singularity of Ugolino's Dantean Space here is an echo of so many others' in the Inferno. We can look back to the first true sufferers on Dante's journey, the meeting with Francesca, who is conjoined with her lover Paolo. Because of their crime of adultery they are now condemned for eternity to be buffeted together by winds through their circle, never having respite or comfort or the pleasures of their love. But of course their love is still on some level bonding them together, less a consolation than a constant reminder of their crime. For such characters as Ugolino and Francesca, each space is in this sense custom-built, a specifically-tailored expression or extrusion of this particular person's own character flaw and history.

And as always, the crime is memorialized and made living and manifest here in physical form. Air, water, ice, flame, space, sepulchres, mountains, rivers, plains, walls; across the Inferno, emotions are made manifest not just in action but also in architecture, nature, geological elements, and in the local suspension of normal physical laws. One's personal history and one's inner experience determine the peculiar configuration and constitution of your own personal Dantean Space.

The core of his dramatic architecture where the internal is externalised in so many ways can be seen in Dante's subtly complex use of the word "Dentro", a word which, since it can mean "within" or "inside" or "in," can have both a physical and an emotional meaning. Throughout the *Commedia* Dante artfully combines the two uses, a clever and crucial conflation that lies at the heart of Dantean Space. And he goes further; while "Dentro" is used in ways that suggest both being inside one's own interiority and emotions and also being inside a setting, inside something concrete, be it the boat of a demon, a ditch in Hell, or a pit of ooze, it is also used to describe abstract or metaphorical things such as being inside a secret or the sea of God's love. And of course, sometimes it is describing a situation or object that is somehow more than one of these things at the same time.

For example consider Canto 7, line 9. Here Virgil says to Plutus the demon: "consuma *dentro* te con la tua rabbia." This can be translated as "May Fury Gnaw You *Inwardly* Away!" In this case Dentro is clearly adverbially describing an *interior* space that is occupied and gradually shaped by one's grievances, emotions, and furies.

But that interior space is often also the architect of one's new-found surroundings in the city of Hell. We see this has happened for example to the Sullen, as they bewail in Canto 7, line 123: "Tristi fummo ne l'aere dolce che dal sol s'allegra, portando *dentro* acciioso fummo: or ci attritiam ne la belletta negra." "We... nursed *within* ourselves sullen fumes, and (thus) come to misery in this black ooze." Here the sullen fumes of these sullen people were grown in an interior space and then as a result they directly–almost cancerously–created the literal miasmatic site, the physical space, of the black ooze that the Sullen are now immersed in.[1] Psychology produces–and continues to be manifest within–the physical Architecture of Dis; here the denizens are all trapped within themselves.

We see this elision of psychic, moral and physical space throughout the *Commedia*, a triple meaning here in the uses of Dentro, a spatial, emotional and subjective elision that is the core of the sublime nature of Dante's spaces. These spaces, while dictated by God's will, are each clearly character-based and character-specific; it is a uniquely dramatic architecture in that one's personal drama is then made manifest in one's Architecture. Whether found in Hell, Purgatory or Heaven, each punishing or purgatorial or celebratory space is somehow composed in its very substance by the actions and emotions, the crimes, vices, loves, hopes, dreams, regrets, hatreds and flaws of the inhabitant.[2]

---

[1] In fact, in the case of the prostrate Capaneus, his emotions themselves are the actual pure cage of Hell in which he is condemned:

> "Oh Capaneus! Pride yet uninterred!
> This punishment, in consequence, is yours.
> No agony, except your own great rage,
> would serve as proper answer to your ire."
> Virgil, Canto 14, lines 62-66.
> (Translated by Robin Kirkpatrick.)

[2] *Dentro* is also found in slightly different contexts in other places in the poem. For example, in line 21 of Canto 3, Dante describes the journey Virgil takes him on through Hell as: "Mi mise *dentro* a le segrete cose." This might be translated as "(he) put me into secret things," an interestingly physical way to describe the truths, stories and spaces of the Commedia. The very metaphorical and spiritual goal for the trip is itself given a physical and Spatial dimension.

It is crucial to see the deeply tactile and physicalized relationships–his imaginative production design, setting and architecture–that Dante specifically constructs in order to intensify the character's dramatic bond with his Space. Immersion, encasement and bodily transformations are constants throughout the Cantos; these are places where characters are encased in searing metal robes or burning sepulchres, knocked into mountainsides by constant buffeting winds, or immersed in water, ice, boiling blood, ooze, a raw flame, a blinding angelic light or an all-encompassing sea-like love.

But of course what is crucial in Dante is the dramatic relationship of this osmosis; the logic of the boundary and loss of boundary, of the heat or cold or piercing or constriction, is governed by a *judicial* and *Aristotelian* logic; it is not simply the *spectacular* logic of horror that we see displayed so gruesomely in so many slasher horror movies. In the *Commedia* there is always a biographical and moral reason for what happens to you; people here truly stew in their own juices, whether that juice is bitter or sweet, acid or balm, and we can then deeply taste this juice; whether through their own speech or by being described by another, their poetic presence is entirely about how their personal history, flaws, pains, fears, loves and conscience-wracked suffering has all led them to this specific place. Dante's gift is to marry punishment and reward with both an outward life of one's past actions and an active inward life of emotions. It is all ruled by the cause-effect logic of Aristotelian drama made physically manifest in architecture.

And so we define Dantean Space as *a kind of projective space in which a character's emotions, passions and character history become dramatically manifest in the physical, architectural and meteorological form of his or her environment. Thus a Dantean Space is like a cave onto the surfaces of which the intimate inner emotional conflicts, imaginings, fears and hopes of the character are being directly projected as a kind of screen.* There is a form of egotism and individualism in Dante's Spaces; here the person is entirely surrounded by himself.

Glancing back now on our four-part taxonomy of Aristotelian spaces, we can see that these spaces of the *Divina Commedia* wonderfully marry a character's External Antagonists or Goals with his or her Inner Antagonisms and Inner Goals, and Dante does this through physicalizing each person's world in specific detailed ways that are psychological, ethical and biographical. This is the specific beauty, horror and drama of

Dantean Architecture, the Space "within" that is also enclosing and embracing.[3]

4. Projective space in cinema

Such a device is clearly present in Cinema. For example, in the *Aliens* movies, where Ripley's Alien spaces are clearly complex Dantean spaces.

As has been pointed out many times in film criticism,[4] Ripley is always in one of two kinds of architectural space. Either they are safe spaces, which are always well-lit, rectangular, usually sterile white spaces, or she is in *dangerous* spaces: dark, organic, rounded, dripping, blackened, intenstine-like spaces that are controlled, occupied and ecologically transformed by the fearsome Aliens.

Over the course of the story, as Ripley increasingly struggles to master her own horrible memories and nightmares of being impregnated and having an alien burst out of her, as she takes charge of the squad of marines and grows stronger as a character, she also moves from the safer, sterile spaces to spaces that are darker, more organic, and more dangerous. Finally, in the penultimate scene when she has to go into the heart of the black dripping, living nest of the Alien mother to retrieve her own daughter-figure, she finally faces her own fears of birth, death and bodily violation. In her final triumphant moments, having vanquished both the alien mother and her own fears, Ripley can at last put her child and herself to bed, alone in a white sterile safe berth in the escaping spaceship.

We might call Ripley's problem an Architecturally-projected conflict, as both the world antagonist and the soul antagonist are manifest there in the architectural spaces through which the film moves. Consider the scene and the space where Ripley is fighting the mother of the Aliens as she also wrestles with her own fear and desire to be a mother. Ripley has an inner and an outer antagonist here, and in the end, after a very long struggle, she

---

[3] There are earlier echoes of this kind of structure in Dante's precursors: elements exist in Sophocles, Euripides and Ovid, for example–and there is a somewhat parallel project in Prudentius' Psychomachia. Still we can comfortably assert that only Dante truly realized, refined and excelled in this kind of poetic justice, and he is by far the greatest inspiration for the emotional/architectural spaces in cinema today.
[4] i.e. Creed, 1990, p. 129.

triumphs over both. The city where Ripley faces her monsters is serving two narrative purposes at once; it is a richly layered Dantean space.

Think of other cases of cinematic plague-infested cities besides *Aliens* that have a similar deep structure to their Dantean Space. Consider, for example, *The Third Man* and *Blade Runner*; their core conflicts are also projected onto their richly realized phantasmagorical cities. It is interesting that all three of these films are plague narratives set in plague cities where the Protagonist, like Oedipus, tries to lift the curse from the oppressed denizens of the cursed city. In all of these films, the Plague is an outer antagonist, but in all three narratives the plague holds another dimension; it is hinted that these protagonists, like Oedipus, are also battling something in themselves at the same time, some aspect of their character that they cannot accept, and that for this reason they are linked to the plague itself on a personal level. Somehow they share some aspect of responsibility for this plague and must find redemption in a specific final choice that will somehow permanently change their lives. Thus in all three films the cities and spaces are functioning as both Outer and Inner Antagonists; as a result these are all rich Dantean Spaces.

I would argue that this revolutionary craft nature is not an accident but is connected to the nature of Dantean Space in cinema. When Dante brought physical and visual dimensions to his intimate portraits of Ugolino, Francesca and so many others, he showed artists how to use the craft of Spectacle to visually and dramatically express. This ability to mirror the intimate in the physical and visual is an especially powerful tool for all art but especially for cinema, where so much effort goes into creating a dramatic bond between viewer and the subjectively-experienced struggles of a protagonist.

For this reason Cinema offers perhaps the greatest potential in realising Dante's invention. Practically speaking, on films such as *Aliens* or *Amélie* that feature Dantean Space, all of the craft departments must dive more deeply into the character's psychology and manifest it, working outside the more usual task of creating plausible, natural backdrops and settings that normally, and invisibly, support the dramas of ordinary DisPassionate and Dramatic Spaces.

In fact, perhaps it is actually *because* these stories require the cinematic elaboration of Dantean Space that they are among the highest examples in our tradition of "Total Cinema," that form of filmmaking where all of the craft departments are being employed in ground-breaking, revolutionary ways. Leaving the frame of standardized Hollywood realism, they can mine more adventurous Romantic, Expressionistic, Gothic and other aesthetics to express the core Emotional conflict in their particular realm, opening a rich new expressive range of craft-choices. After all, Dantean Space singularly offers an art form of *subjectivity*, a way out of mimetic realism, a form of storytelling based on the Character's POV, one where we enter the sensory point of view of the main protagonist and "see through their eyes," "walk in their shoes" and "enter their heart, mind and imagination." Such films can construct not just a plausible setting but rather any setting that reflects the character's inner dramatic tension.

This explains why Cinema is often at its most powerful when it is expressing Dantean Space. Cinema is always married to Spectacle, but when Spectacle is also a deep dramatic expression, element and agent of the intimate Inner Conflicts, hopes and dreams of the main characters, then Cinema is raised to a new level of intensity. A film then becomes a specifically Dantean Spectacle, one serving a narrative drive, and thus becomes much more powerful than conventional cinema's Spectacle, where the mise-en-scène generally serves only either as the mere backdrop of DisPassionate Space (in the industry, cases of this form of Production Design and Art Direction are called "eye-candy") or as the visually-arresting arena of the drama–the cases we are calling Dramatic Space.

We are suggesting then that Dantean Space is at the core of so much of contemporary cinema. One genre that has made best use of it is Horror, as we have hinted at here, but in future work we will return to this again in essays on Purgatorial and Heavenly cities to see how the same kind of Dantean Space is used in Romantic Comedies and Dramas to express a character's hope, love and Virtuous behaviour. For now we will only point out that Dante's invention is also deployed in Purgatorial and Heavenly spaces.

Conclusion

In this short paper we have tried to historicise the narrative tradition of Spaces by tracing out the different dramatic techniques for Architecting cities in stories and cinema. Here I have offered a taxonomy and set of definitions that, methodologically speaking, I think has the double merit of being both simple/familiar and used by the very practitioners of cinema today. I then try to use all the three tools to explain the relationship in a film or story between a protagonist and any setting he or she might be strolling or racing or fighting through. I have also offered a definition of Dantean Space, a category of aesthetic invention that I argue starts largely with Dante but exists in much of contemporary cinema.

Of course, deconstructing the tradition using its own aesthetic categories will next lead us directly to the grand interrogation of such terms as Identity, Self-identity, Interiority, Subjectivity and the Bourgeois Subject. If this account of Dantean Space is correct in its broad strokes, then we must now ask the questions about Subjectivity that have been posed in Sociology, Anthropology, Cultural Theory and Post-Colonial Studies, that great grand project brilliantly realized in the work of, for example, Jacob Burckhardt, The Frankfurt School, Edward Said, Stephen Greenblatt, Terry Eagleton and so many others.

In other words, after giving this account of Dantean Space we might now ask a few questions. First of all, what is the impact on subjectivities of stories that unfold in Interior Space in this way? Do such stories simply deepen or reveal Interior Space or–as Burckhardt and Greenblatt might claim–do they actually somehow create, engender or unfold it? Do they thus also create in some sense the psychological and aesthetic machinery of subjectivity?

Secondly, how has this unfolding dialectic of Storytelling technique actually affected and shaped Identities and societies? Has it had a role in the violent collisions of different aesthetic, political and economic cultures?

Thirdly, what are the implications of telling a story that features Dantean Space in a culture that has no prior aesthetic tradition of this kind?

In short, has Dante actually architected not only our cinematic cities but even our very subjectivity, the very conscious minds we now possess? And has his invention helped create actual Heavenly, Purgatorial or Hellish Spaces on earth?

References

Alighieri, D. (2006). *The Divine Comedy 1: Inferno*. Translated and edited by Robin Kirkpatrick. London: Penguin.
—. (2003). *The Divine Comedy 2: Purgatorio*. Translated and edited by Robert M. Durling. London: Penguin.
—. (2008). *The Divine Comedy 3: Paradiso*. Translated and edited by Robin Kirkpatrick. London: Penguin.
Aristotle, (1961). *Poetics*. New York: Hill and Wang.
Burckhardt, J. (1860/1965). *The Civilization of the Renaissance in Italy*. London: Phaidon Books.
Creed, B. (1990). Alien and the Monstrous-Feminine. In A. Kuhn (Ed.), *Alien Zone: Cultural Theory and Contemporary Science Fiction Cinema* (p. 129). New York: Verso Books.
Cunningham, D. A. (2003). Stalling Zion: Hegemony through Racial Conflict in The Matrix. *Rhyzomes*, 7, Fall.
Egri, L. (2007). *The Art Of Dramatic Writing*. Rockville, MD: Wildside Press.
Field, S. (2005). *Screenplay: The Foundations of Screenwriting*. (Revised Edition). New York: Delta Press.
Gardener, E. (Ed.). (2008). *Visions of Hell and Heaven Before Dante*. New York: Italica Press.
Greenblatt, S. (1990). *Learning To Curse: Essays in Early Modern Culture*. London and New York: Routledge.
—. (1980). *Renaissance Self-Fashioning: From More To Shakespeare*. Chicago: University of Chicago Press.
Lefebre, H. (1992). *The Production Of Space*. Oxford: Wiley-Blackwell.
Ovid, (1995). *Metamorphosis*. Wilmington, MA: Mariner Books.

Film Citation List

Aliens. Dir. Ridley Scott. Perf. Sigourney Weaver, Michael Biehn, Carrie Henn. Twentieth Century Fox Film Corporation, Brandywine Productions, SLM Production Group, 1986.
Amélie. Dir. Jean-Pierre Jeunet. Perf. Audrey Tatou, Mathieu Kassovitz, Rufus. Claudie Ossard Productions, UGC, Victoires Productions, 2001.
Escape From New York. Dir. John Carpenter. Perf. Kurt Russell, Ernest Borgnine. AVCO Embassy Pictures, International Film Investors, 1981.

Fight Club. Dir. David Fincher. Perf. Brad Pitt, Edward Norton, Helena Bonham Carter. Fox 2000 Pictures, Regency Enterprises, Linsom Films, 1999.

The Graduate. Dir. Mike Nichols. Perf. Dustin Hoffman, Anne Bancroft, Katherine Ross. Lawrence Turman, Embassy Pictures Corporation, 1967.

Inception. Dir. Christopher Nolan. Perf. Leonardo Di Caprio, Joseph Gordon-Levitt, Ellen Page. Warner Brothers Pictures, 2010.

The King's Speech Dir. Tom Hooper. Perf. Colin Firth, Geoffrey Rush, Helena Bonham Carter. The Weinstein Company, UK Film Council, Momentum Pictures, 2010.

The Matrix. Dir. Andy Wachowski, Lana Wachowski. Warner Bros. Pictures, Village Roadshow Pictures, 1999.

Midnight Cowboy. Dir. John Schlesinger. Perf. Dustin Hoffman, Jon Voight. Florin Productions, Jerome Hellman Productions, 1969.

Los Olvidados. Dir. Luis Buñuel. Perf. Alfonso Mejia, Roberto Cobo. Ultramar Films, 1950.

Rome, Open City. Dir. Roberto Rossellini. Perf. Anna Magnani, Aldo Fabrizi, Marcello Pagliero. Excelsa Films, 1945.

Salaam Bombay. Dir. Mira Nair. Perf. Shafiq Syed, Hansa Vithal, Chanda Sharma. Cadrage, Channel 4 Films, 1988.

Slumdog Millionaire. Dir. Danny Boyle. Perf. Dev Patel, Freida Pinto. Celador Films, Film4, Pathe Pictures International, 2008.

The Third Man. Dir. Carol Reed. Perf. Orson Welles, Joseph Cotton, Alida Valli. London Film Production, British Lion Film Corporation, 1949.

CONTRIBUTORS

Giovanni Caruso
Is a Ph.D. Candidate in Audiovisual Studies at the University of Udine. He investigates cultural connections and implications of in-game user generated contents with a focus on design and use. He is editor and co-founder of *G|A|M|E – Game as Art, Media, Entertainment – The Italian Journal of Game Studies* and co-editor of *Piracy Effect,* a forthcoming book on digital piracy.

Mariana Ciancia
Is a Ph.D. Student in Design and Multimedia Communication at the Department of Industrial Design, Art, Communication and Fashion (INDACO), School of Design, Milan Polytechnic. Her research activity deals with media convergence, participative and collective culture, focusing on Transmedia Communication Systems.

Patrick J. Coppock
Is Adjunct Professor in "Philosophy of Languages" at the School of Communication and Economic Sciences at the University of Modena and Reggio Emilia, Italy. He has taught introductory courses in semiotics, new media and communication science. Since 2008, he has been coordinating a MA/MSc level research seminar conducted in English on the Philosophy and Theory of Social and Ludic Media. He is a founder member of *gamephilosophy.org* and coordinator for the *Game Philosophy Initiative @UNIMORE*.

Amedeo D'Adamo
Was the Founding Dean and then President of the Los Angeles Film School, a floating graduate film-crafts conservatory in Hollywood, CA. He is currently a Visiting Professor in the Cinéma MA at the University of Switzerland (It), and teaches Directing at the Catholic University, Italy. He was also the Scientific Director of the Puglia Experience, an innovative fellowship program for professional screenwriters funded by the EU which he ran for the Apulian Film Commission from 2008 to 2011. Amedeo also co-founded The Traveling Film School, a non-profit organization that provides free film and theater training to children in underdeveloped

regions; it has built small tuition-free film-schools in Cameroon and Sardinia. He wrote and, with his wife Nevina Satta, co-directed and co-produced the UNICEF-endorsed feature *The Tree Of Ghibet*, a drama about the exploitation of street children (Rome F.F. '08, Miami '08, Austin '08, Torino '07). He holds an MFA in Screenwriting from Columbia University's Graduate Film Program.

Miriam De Rosa
Earned her Ph.D. in Culture of Communications at the Catholic University of the Sacred Heart of Milan, where she currently holds the seminar "Culture of the Audiovisual" and collaborates to the activities of the Media and Performing Arts Department. Her approach is based on a Visual and Film Studies background, and her researches explore mainly the relationship among subject, space and filmic device, focusing on contemporary forms of cinema. She is the editor of the project *Screencity* and a member of NECS.

Manuela Farinosi
Is a Post-Doctoral Researcher in the Department of Human Sciences at the University of Udine (Italy). She teaches Economic Sociology and collaborates on the courses of Sociology of Communication and Culture and Theories and Techniques of New Media. Her main research interests are focused on social and cultural aspects of digital technologies, with particular attention to social media and their impact on public and private spheres. She has been Visiting Scholar at Alberta University (Canada) and she is a member of the COST Action FP1104.

Riccardo Fassone
Is a Ph.D. Candidate in Media History at the University of Turin, Italy. His dissertation is focused on the notions of ending and closure in video games. His research interests are video game theory, genre theory and media history. In 2012 he was Visiting Scholar at the Georgia Institute of Technology. He has published several articles and papers on video games and popular cinema. He is the editor and co-founder of *G|A|M|E – Game as Art, Media, Entertainment – The Italian Journal of Game Studies*.

Katalin Fehér
Is a Research Fellow at the Research Centre of the Budapest Business School (Hungary). Her research interests are trends in digital and virtual media. Her current focus of interest is digital culture, platforms and online

patterns of contents, social media, changing scales, digital identity and online brands.

Gabriele Ferri
Holds a Ph.D. in Semiotics from SUM - Istituto Italiano di Scienze Umane (Florence, Italy) and the University of Bologna, where he teaches the Computer Game Design and Analysis Lab. He is a freelance Urban Game designer for several associations and public bodies and he is a member of the editorial committee of *G|A|M|E – The Italian Journal of Game Studies*.

Chiara Giaccardi
Ph.D. in Social Sciences (University of Kent, UK), is Full Professor of Media Sociology and Anthropology at the Catholic University of the Sacred Heart of Milan, where she teaches Media Sociology and Anthropology and collaborates with the Centre for the Anthropology of Religion and Cultural Change. She is the director of the academic Journal *Comunicazioni Sociali*; founding member and chair of the ECREA Temporary Working Group 'Media & The City'. Her research interest focuses on the cultural impact of communication technologies on identity and intercultural relationships, web and post-secularism, the relationship between ICTs and urban space.

Sami Kolamo
Is trained both in human geography (University of Oulu) and media studies (University of Tampere). He is currently Researcher at the School of Communication, Media and Theatre, University of Tampere. His main research interests deal with sports mega-events, mediated urban environments and fan activities. He is in the board of the Finnish Society for Cinema Studies and is a member of ECREA's Media & the City working group. His Ph.D. dissertation on football media spectacles will be published at the beginning of next year.

Walter Mattana
Is a Ph.D. Student in Design and Multimedia Communication at the Department of Industrial Design, Art, Communication and Fashion (INDACO), School of Design, Milan Polytechnic. His research activity deals with the field of audiovisual storytelling, focusing on the culture and languages of cinema and film analysis.

Gonca Noyan
Is a Ph.D. Candidate in communication at the University of Illinois at Chicago (UIC). She has taught Media Communication at UIC. She holds an M.A. degree in public diplomacy from the University of Southern California (USC). She is co-founder of *Soft Power TR*, an ad-hoc group of academics and communication consultants providing strategic insights and policy recommendations on public diplomacy, soft power and city branding. She is a Fulbright Scholar and Young Atlanticist at the Atlantic Council. Her research interests include globalization, public space and multiculturalism in post-conflict societies.

Christian Oggolder
Is Senior Scientist at the Institute for Comparative Media and Communication Studies (CMC) of the Austrian Academy of Sciences and the Alpen-Adria-University Klagenfurt. He holds a PhD in History from the University of Vienna and a Master in New Media Management from the Danube University Krems. His major research interests are media system analysis, media history, internet, and the social web. Currently he is involved in the projects 'The Social Web and Changing Media Systems' and 'World Internet Project'.

Seija Ridell
Works as Professor at the School of Communication, Media and Theatre of the University of Tampere (Finland). Her areas of expertise include news narratology, social semiotic theory of genre, cultural audience research and web studies. Her most recent research concerns media and technology related modes of (public) action in diverse spatial contexts.

Mauro Salvador
Is a Ph.D. Candidate in Culture of Communications at the Catholic University of the Sacred Heart of Milan. His research project deals with the connection between video games, other media and reality, with gamification processes and with the spreading influence of games in media environments. He achieved a M.A. in "Cinema, television and multimedia production" at the University of Bologna with a thesis about contemporary cinema audiences and their mixing with other media users. He is the editor and co-founder of *G|A|M|E – Game as Art, Media, Entertainment – The Italian Journal of Game Studies*. He is interested in games, contemporary cinema, new media and all that concerns their relationships.

Gabriella Sandstig

Has a Ph.D. in Journalism and Mass Communication. She is Senior Lecturer at the Department of Journalism, Media and Communication of the University of Gothenburg and Researcher at the Department of Political Science of the University of Gothenburg. She is a media effects researcher with focus on the importance of contextual factors like urban space.

Barbara Scifo

Ph.D., is Senior Lecturer in Sociology of Culture and Communication at the Catholic University of the Sacred Heart of Milan, where she teaches Languages and Instruments of Communication. She is a Senior Researcher at *OssCom - Centre for media and communication research*. Her main research interest is the social shaping of ICTs, especially the mobile media and internet. She has recently paid special attention to young people's crossmedia practices and audience research in the social media era. She is currently engaged in the study of the "domestication" of the mobile internet and of the smartphone by children and young people and in a research project on social network, identity and social relations in the Italian context.

Satomi Sugiyama,

Ph.D., is Associate Professor of Communication and Media Studies at Franklin College Switzerland. She has been conducting research on the way young people perceive and use mobile communication technologies in various cultural contexts, particularly in Japan. She completed her Ph.D. in Communication at Rutgers University in 2006. She was awarded the MacArthur and National Endowment for the Humanities pre-doctoral fellowships at Colgate University. She has organised an international exploratory workshop on social robots and emotion upon receiving a grant from the Swiss National Science Foundation in 2011, and is currently examining the boundary between humans and mobile technologies.

Moira Sweeney

Is an artist and filmmaker. Her lens-based doctoral study at the Centre for Transcultural Research and Media Practice, Dublin Institute of Technology is titled *Dublin Dockers - Visualising a Changing Community*. Among her current broadcasts/exhibitions/presentations are: "Aniar Aduaidh": Director, Music Documentary Series for RTÉ, 2013; "Stevedoring Stories": Multi Media Installation, exhibited at Dublin Tall Ships, 2012; "Space and the Geographical Imagination on the Dublin

Docklands": Poster Presentation, Social Media and Global Voices, ECREA, Istanbul, 2012 and Oral Presentation, Media and the City International Workshop, Catholic University of the Sacred Heart of Milan, 2012.

Matteo Tarantino

Is Postdoctoral Fellow at the Catholic University of the Sacred Heart of Milan, where he lectures in Media & Social Networks and collaborates with the Centre for the Anthropology of Religion and Cultural Change. His research interests focus on the cultural impact of communication technologies (especially in non-western contexts) and on the relationship between ICTs and urban space. He is a founding member of the ECREA Temporary Working Group "Media & The City". He has collaborated with the Hong Kong Polytechnic and the Communications University of China.

Federica Timeto

Is a Ph.D. candidate in "Computing, Communications and Electronics" at the University of Plymouth (Planetary Collegium), Uk/Milan Node (Milan, Nuova Accademia di Belle Arti). Her research programme is called "Feminism, Postcoloniality and New Media Theory and Practice" and focuses on feminist artistic practices and the new media in a postcolonial perspective. Her interests focus on art history criticism and feminist aesthetics, feminism and new technologies, postcolonial studies, cultural and visual studies. She is Lecturer of Arts Criticism History and Methodology and Phenomenology of Contemporary Arts at the Academy of Fine Arts Abadir (Palermo), and teaches Art History in High School.

Simone Tosoni

Is Assistant Professor at the Catholic University of the Sacred Heart of Milan, where he lectures in Media & Social Networks and Digital Media and Social Processes, and collaborates with the Centre for the Anthropology of Religion and Cultural Change. His research interests focus on the cultural impact of communication technologies, identity and ICTs, and on the relationship between ICTs and urban space. He is a founding member of the ECREA Temporary Working Group "Media & The City".

Emiliano Treré

Has a Ph.D. in Multimedia Communication from the University of Udine (Italy) and is currently appointed as Associate Professor of Digital Media and Social Movements at the Department of Communication and

Journalism in the Faculty of Political and Social Sciences of the Autonomous University of Querétaro (Mexico). He teaches courses on digital and community media, social movements, cybercultures and qualitative methodologies. He has been Visiting Scholar at Dartmouth College (US), Erfurt University (Germany) and Complutense University (Spain). His work has appeared in New Media & Society, the International Journal of Communication and in edited books in Spanish, English and Italian. He is currently working on a project concerning the mapping of Mexican activists' media practices.